RAND NATIONAL DEFENSE RESEARCH INSTITUTE

T0302879

Reserve Component General and Flag Officers

A Review of Requirements and Authorized Strength

Lisa M. Harrington, Igor Mikolic-Torreira, Geoffrey McGovern, Michael J. Mazarr, Peter Schirmer, Keith Gierlack, Joslyn Hemler, Jonathan Welch

Prepared for the Office of the Secretary of Defense

For more information on this publication, visit www.rand.org/t/RR1156

Library of Congress Cataloging-in-Publication Data is available for this publication.
ISBN: 978-0-8330-9098-0

Published by the RAND Corporation, Santa Monica, Calif.

Preface

The Fiscal Year 2014 National Defense Authorization Act directed the Secretary of Defense to conduct a review of requirements for reserve component general and flag officers (RC G/FOs) in an active status. The Office of the Assistant Secretary of Defense for Reserve Affairs (OASD/RA) asked RAND to assist with this assessment; this report provides OASD/RA with information relevant to that request. OASD/RA asked RAND to explore (1) whether the authorized strengths for RC G/FOs, as provided in statute, are sufficient for the effective management and leadership of the reserve component; (2) whether the services are complying with statutory limits on the number of general and flag officers; (3) whether efficiencies can be gained from downgrading or eliminating RC G/FO positions or from converting them to senior civilian positions; and (4) the extent to which professional development of RC G/FOs affects requirements.

Our research is based on information gleaned from a number of sources: policy and legal documents guiding the authorizations and management of general and flag officers and positions, previous studies on the requirements for general and flag officers, interviews with subject-matter experts and stakeholders, and analysis of data collected from the military services on all RC G/FOs. The results of our research indicate that the military departments are complying with statutory limits on the number and grade of RC G/FOs. There are, however, some RC G/FO requirements that the reserve components cannot fill because of these limits. In addition, we have identified a small number of cases where the requirement for an RC G/FO is unclear.

This research was sponsored by the Office of the Under Secretary of Defense for Personnel and Readiness. It was conducted within the Forces and Resources Policy Center of the RAND National Defense Research Institute, a federally funded research and development center sponsored by the Office of the Secretary of Defense, the Joint Staff, the Unified Combatant Commands, the Navy, the U.S. Marine Corps, the defense agencies, and the defense Intelligence Community. Comments on this report are welcome and may be addressed to Lisa Harrington at lharring@rand.org.

For more information on the RAND Forces and Resources Policy Center, see http://www.rand.org/nsrd/ndri/centers/frp.html or contact the director (contact information is provided on the web page).

Contents

Figures and Tables

Figures

Tables

Summary

The optimal balance between force size and the number of general and flag officers has been a question of interest in military personnel planning for decades. Increased use of the reserve component, particularly since September 11, 2001, and an expectation that this role will continue and possibly increase in the future has renewed interest in the question of the appropriate number of reserve component general and flag officers (RC G/FOs). Congressional interest in this question prompted legislation in the Fiscal Year 2014 National Defense Authorization Act, which called for the Secretary of Defense to conduct a review of "general officer and flag officer requirements for members of the reserve component in an active status . . . to ensure that the authorized strengths provided in Section 12004 of Title 10, United States Code are based on an objective requirements process," are sufficient, and reflect any efficiencies that can be gained by eliminating or downgrading unneeded positions.

The Office of the Assistant Secretary of Defense for Reserve Affairs (OASD/RA) asked RAND's National Defense Research Institute to assist with this assessment and answer the following research questions:

- Does the current system of authorized strength, including exemptions, applicable to RC G/FOs provide for a sufficient quantity of senior officers?
- Are the military departments appropriately managing and accounting for RC G/FO requirements, authorizations, authorized strength, and inventory?
- Are there candidate positions that should be considered for elimination, downgrading to a lower rank, or conversion to senior civilian positions?
- To what extent should the need to provide developmental experiences for RC G/FOs drive requirements?

A challenging aspect in accomplishing this research was the interrelated aspect of general and flag officer management and accounting. Several terms need to be clarified:

- *Requirements* are positions that the services and Joint Staff designate for a general or flag officer of a particular grade; these positions are usually listed as such in a manpower document.

- *Authorizations* are positions (requirements) that the secretary or chief of the service (for service authorizations) or the Chairman of the Joint Chiefs of Staff (for joint authorizations) has decided to fill with a general or flag officer.
- *Authorized strength* is the maximum number of general or flag officers established in law (for the active component, the reserve component, and each military department); this includes both the basic authorized strength and additional authorized strength in the form of exemptions.
- *General or flag officer inventory* is the number of individuals who can be assigned to fill general or flag officer authorized positions.

What is important to understand is that requirements and authorizations reflect numbers of positions, authorized strength is the number of people allowed to fill those positions, and inventory is the population of people available to fill the positions—and these numbers may or may not be equal. In fact, requirements are generally greater than authorizations, and authorizations are generally less than or equal to authorized strength. This is because total requirements are not constrained by authorized strength limits, whereas filling authorizations with a general or flag officer is constrained by the availability of authorized strength.

Are Authorized Strength and Exemptions Sufficient?

The services report that RC G/FO authorized strength and exemptions are sufficient to provide effective management and leadership of the reserve component and to fulfill the varied roles in which RC G/FOs serve. The exemptions, in particular, are necessary to meet the full range of RC G/FO requirements for such roles as performing joint duty, providing National Guard leadership at the state level, and filling skill gaps for general and flag officers in the active component. While there may be particular requirements that the management offices are unable to fill from time to time, there is no consistent type or set of requirements that they are unable to fill because of limits on authorized strength. As a result, we recommend:

- *Congress and the Department of Defense (DoD) maintain the current system of basic authorized strength plus individual exemptions.* While the general and flag officer management offices reported that the structure of statutory strength and exemptions is complex and difficult to manage, we did not find sufficient evidence to support changes to statute. The current categories of exemptions provide for necessary full-time, developmental, and leadership roles for RC G/FOs and give the services flexibility to use reserve component officers when responding to changing mission needs. A single pool combining basic authorized strength and exemptions would remove flexibility that is essential to effective use of these officers.

Are the Services Appropriately Managing and Accounting for RC G/FO Positions?

The services' current requirements and authorizations processes are objective and able to meet service needs but would benefit from a systematic, comprehensive review process. The reserve components and joint organizations develop job descriptions for positions that should be filled by a general or flag officer that take into consideration a number of characteristics: the nature of the position's duties, magnitude of responsibility, span of control and scope of the resources managed, significance of actions and decisions, importance of the mission to national security, and others.

Job descriptions notwithstanding, no *formal*, consistent classification standards or guidelines exist that the services can use and reference. However, all the general and flag officer management offices did report that the military judgment of service and joint senior leaders is the single most important consideration when decisions are made to create or validate a general or flag officer position.

RAND researchers found that differences in individual service philosophies on the appropriate roles for RC G/FOs could result in inconsistencies when designating these positions. For example, positions with similar responsibilities can be assigned to individuals at different grades depending on the service. As a result, the services could benefit from a process for reviewing and validating RC G/FO positions that facilitates better coordination throughout DoD.

The services and the Joint Staff have appropriate and accurate methods for ensuring they comply with statutory limitations on the number of general and flag officers—including both basic authorized strength and the many exemptions. But these methods are inconsistent and it took significant time and effort for RAND researchers to evaluate the methods and ensure that the services were interpreting and adhering to statutory limitations. This is particularly true of the exemptions to the basic authorized strength.

Although these management systems generally work, we found opportunities for improvement and recommend that the Office of the Under Secretary of Defense for Personnel and Readiness (OUSD[P&R]):

- *Develop guidelines for periodically reviewing and validating RC G/FO requirements.* Previous studies of general and flag officer requirements (for the active and reserve components) emphasize the difficulties in determining how positions requiring general and flag officers are designated or how many officers are needed. We recommend that OUSD(P&R) develop and release guidelines for evaluating and validating RC G/FO positions. The guidelines should encourage consistency across service and joint positions while recognizing individual mission needs of the services, Joint Chiefs of Staff, Office of the Secretary of Defense, and combatant commands. They should also encourage a formal, visible process for creating or validating general and flag officer positions, such as the one currently being used by the Air Force Reserve.

- *Institute methods for DoD-wide tracking of RC G/FO authorized strengths.* We recommend that OUSD(P&R) institute a department-wide system for tracking the RC G/FOs provided for under the basic authorization (422 total under 10 U.S.C. 12004) and the individual categories of exemptions. With a common tracking system, OUSD(P&R) can more easily ensure that the number of RC G/FOs does not exceed statutory limitation and more accurately track the use of exemptions over time. A common tracking system should foster increased collaboration between active and reserve component general and flag officer management offices to share best practices in making best use of authorizations and exemptions.
- *Resist attempts to link RC G/FO limits to total service or component end strength.* The size of the RC G/FO corps should not be tied to fluctuations in end strength or to the use of RC G/FOs in times of war or national emergency. Basing general and flag officer authorized strength on reserve component end strength could limit the use of RC G/FOs where they provide reserve component expertise, gain experience, and fill in for the active component, but do not lead large numbers of troops. We recommend that OUSD(P&R), the military departments, and the Joint Staff oppose limitations to RC G/FO authorized strengths based on end strength, as such limitations would be counterproductive.

Can Positions Be Eliminated, Downgraded, or Converted?

There are potential opportunities to gain efficiencies by eliminating or downgrading a limited number of RC G/FO requirements. After evaluating 634 positions, RAND researchers identified between 27 and 95 positions that are possible candidates for elimination or downgrade—depending on how stringently the factors were applied to existing requirements. But before any action is taken, the military departments should review and revalidate these positions as general and flag officer requirements—since important contextual, qualitative, and subjective factors that may justify the position could not be assessed in RAND's purely analytic process.

We found very few opportunities for converting RC G/FO requirements to senior civilian positions. The biggest impediment is that many RC G/FO positions are part-time positions, making them impractical to fill with civilians. In addition, many RC G/FO positions are valuable for officer development—in fact, 65 percent of the reserve component positions on general staffs are identified as having value for the professional development of RC G/FOs.

RAND researchers did identify one category of positions—mobilization assistants or National Guard assistants—that needs to be reevaluated. These positions have loosely defined peacetime roles that can be generically described as liaisons between their commands and the reserve components, but their roles upon mobilization are not well articulated. These positions constitute one-third of the RC G/FO requirements, not counting the state adjutants general and assistant adjutants general positions. Thus, establishing overarching service policies and standards that address the expected peacetime and mobilization roles of these assistants could help justify them and assuage concerns. Because most of these positions are filled by individual mobilization augmentees (IMAs), department-wide guidance on how to use RC G/FO IMAs and the roles they should play would also help justify these requirements.

RC G/FO requirements that have emerged since the terrorist attacks of September 11, 2001, broadly reflect the needs of the past 15 years. Many of these requirements are closely related to enduring changes in force structure and organizations, but others are more closely tied to operational requirements that continue to evolve. These latter requirements should be reviewed on a regular basis to ensure that limited general and flag officer resources are dedicated to the highest priority missions.

The services and joint organizations can take a number of actions to better justify existing requirements for RC G/FOs. We recommend:

- *The services and joint organizations should review the positions identified in our analysis to ensure these remain valid RC G/FO requirements.* Those that are no longer valid should be considered for elimination or downgraded.
- *OUSD(P&R) should assist the services and the Joint Staff in establishing overarching guidance on how RC G/FO IMAs should be used and the roles they should play.* While flexibility to tailor assignments is essential, there should be a common foundation that establishes the basis for how all the services employ RC G/FO IMAs.
- *OUSD(P&R) should assist the services in establishing guidance regarding the overall nature of positions described as mobilization assistants or National Guard assistants.* There should be a common understanding of the duties and responsibilities associated with such positions both in peacetime and at times of mobilization. Such guidance should assist the services in validating and justifying these positions.
- *The services and joint organizations should review RC G/FO requirements established during the past 15 years.* Some of these requirements are clearly enduring, but others are more closely tied to operational requirements in the U.S. Central Command area of responsibility. As these operations end or evolve, there may be opportunities to free up valuable RC G/FO resources for use elsewhere.

Should Professional Development Considerations Influence Requirements?

The services possess an adequate array of positions to develop a sufficient pool of qualified officers. The services view development as broadening and use assignments to achieve that goal. They have a relatively shared understanding of the assignments necessary to develop most general and flag officers for more-senior leadership, including command at the O-7 level, planning and budgeting, and joint experience. The current slate of general and flag officer positions is generally adequate to support this process.

The most fundamental question we pursued in this aspect of the study was whether the services possess a sufficient array of jobs to create an adequate pool of well-rounded reserve component officers from which to select future leaders. If the services routinely lamented their inability to get officers a certain type of experience, or worried that their requirements were concentrated too heavily in a few areas to generate enough well-rounded leaders, there could be an argument for adjusting the requirements in the name of professional development.

Our research does not suggest that this is the case, however. None of the offices or senior leaders with whom we met, nor our analysis of career paths, suggested any serious gaps in the ability to generate a sufficient number of qualified officers for more-senior positions. There appears to be a sufficient range and diversity of positions to provide the necessary experiences. And just as important, when constraints arise, they are usually a result of limitations in National Guard and reserve assignment processes, such as the location and availability of officers, not the absence of a particular job.

Development, therefore, does not appear to offer clear or specific guidance for judging National Guard and reserve general and flag officer positions. The process of leader development aims to prepare a sufficient number of officers for senior positions, but it should not drive requirements. There are, however, several aspects of development that can and should be taken into account by the services as relates to general and flag officer positions. We recommend the following:

- *Future reductions in general and flag officer requirements or authorizations should not disproportionately affect those positions that are viewed as developmentally crucial.* The services should strive to maintain a good balance of command, joint, headquarters staff, international, combatant command, major command, and other types of positions. The most common and important route to development for senior responsibilities is by broadening the range of assignments that officers can hold. Cuts that fall disproportionately on less "core" or typical command assignments (such as general and flag officers assigned in the joint environment) could have a serious effect on the professional development of general and flag officers.

- *Best practices, ideas, and insights from current leader development efforts in the services should be shared to catalyze further experimentation and innovation in leader development.* While this analysis has been confined to the implications of development for RC G/FO requirements, it has generated broader insights into the current status of officer development programs in the various services. Innovative and promising efforts are under way in many of the reserve components to enhance career-long leader development, both before and after officers are promoted to the general and flag officer ranks.

Final Thoughts

The bottom line is that the process of managing and tracking RC G/FOs is complex and dynamic—driven by a combination of service requirements, authorizations from service and joint leadership to fill positions, authorized strength as defined in statute, and available inventories of officers. Despite the challenges, the services are working within the statutory limits and have relatively solid requirements processes—though these processes could be enhanced to make them more rigorous and consistent. In general, the reserve components have a sufficient number of general and flag officers to meet mission requirements and provide ample opportunity to develop future officers. And the system of authorized strength—with basic authorizations and numerous exemptions—appears sufficient and serves the reserve component well. Eliminating exemptions, by combining them into a single pool with basic authorized strength, would create a rigid system and reduce flexibility that the reserve components need to meet changing requirements and fulfill their varied roles. After examining the validity of current RC G/FO positions, we identified a small number of candidate positions that could, after review by the services, be considered for downgrading or elimination but determined that the vast majority of positions are well justified.

Acknowledgments

The authors would like to express sincere gratitude to all of the interview participants who graciously offered their time and insights—especially the individuals assigned to each of the components' general and flag officer management offices and the general and flag officers who offered their views on the development process. Our research benefited greatly from the expert advice of our principal advisor, Bernard Rostker, and other members of our advisory group: G. Kim Wincup, Daniel Ginsberg, and Greg Schumacher. In addition, the report benefited greatly from the assistance of Barbara Bicksler, who provided her communications and writing expertise on several briefings and this document. We also thank John Winkler, who provided thoughtful guidance on our overall methodology and approach, and Al Robbert for his thorough and thoughtful review of the document.

CHAPTER ONE

Introduction

Following World War II, Congress passed the Officer Personnel Management Act in 1947, which established the number of authorized general and flag officers that could be on active duty. More than a decade later, in 1958, a ceiling on the number of reserve component general and flag officers was established in law (Pub.L. No. 85-861, Section 3218, 1958). Despite these ceilings, the size of the general and flag officer corps for both the active and reserve component has been the subject of numerous congressional hearings and congressionally mandated studies dating as far back as 1966 (Offenhauer, 2007; U.S. Government Accountability Office [GAO],[1] 2004; Conley et al., 2008). Since the end of the Cold War, inquiries have focused on whether the U.S. armed forces have achieved an optimal balance between the size of the force and the number of general and flag officers (see, for example, Offenhauer, 2007), given the series of force expansions and reductions associated with U.S. armed forces' engagement in conflicts around the world. In addition, concerns over the high cost of military personnel and budget reductions have generated increased public scrutiny.

The dramatic increase in the use of the reserve component since the end of the Cold War, especially after the terrorist attacks of September 11, 2001, and the expectation that its role will continue and possibly increase in the future have lent energy to the question of what is the appropriate number of reserve component general and flag officers (RC G/FOs). Some argue that authorized strength is not sufficient given the increased reliance on the reserve forces. Others suggest that the post–Cold War drawdown included a reduction in officer strengths at all ranks in the active component and, thus, RC G/FO positions should also be reduced. Recently, Congress called for yet another review of the RC G/FO corps. Specifically, the Fiscal Year 2014 National Defense Authorization Act (Pub. L. No. 113-66, Section 514, 2013) states that

> The Secretary of Defense shall conduct a review of the general officer and flag officer requirements for members of the reserve component in an active status. The purpose of the review is to ensure that the authorized strengths provided in section

---

[1] *GAO* refers to both the U.S. General Accounting Office and the U.S. Government Accountability Office. The name changed in 2004.

12004 of title 10, United States Code, for reserve general officers and reserve flag officers in an active status—

1. are based on an objective requirements process and are sufficient for the effective management, leadership, and administration of the reserve components;
2. provide a qualified, sufficient pool from which reserve component general and flag officers can continue to be assigned on active duty in joint duty and in-service military positions;
3. reflect a review of the appropriateness and number of exemptions provided by subsections (b), (c), and (d) of section 12004 of title 10, United States Code;
4. reflect the efficiencies that can be achieved through downgrading or elimination of reserve component general or flag officer positions, including through the conversion of certain reserve component general or flag officer positions to senior civilian positions; and
5. are subjected to periodic review, control, and adjustment.

The Office of the Assistant Secretary of Defense for Reserve Affairs (OASD/RA) asked RAND to assist with this assessment by exploring statutory guidance on RC G/FO authorized strengths and the military departments' current methods for determining and managing requirements. OASD/RA also asked RAND to develop a methodology for reviewing RC G/FO requirements and to evaluate how additional factors, such as general and flag officer development, affect requirements. This report provides OASD/RA with information relevant to those requests.

In discussing their perspective on the questions asked by Congress, OASD/RA officials made it clear that the underlying premise of the review was to conduct an objective assessment of the current positions. Thus, the focus of our research was not on validating general and flag officer requirements or on conducting a formal ranking of requirements by level of importance. Instead, the goal was to gain an understanding of RC G/FO requirements, authorizations, and authorized strength in order to develop a process that could be used to periodically evaluate requirements.

Research Questions and Method

In consultation with our sponsor, we identified the following broad questions to guide our efforts:

- Does the current system of authorized strength, including exemptions, applicable to RC G/FOs provide for a sufficient quantity of senior officers?
- Are the military departments appropriately managing and accounting for RC G/FO requirements, authorizations, authorized strength, and inventory?
- Are there candidate positions that should be considered for elimination, downgrading to a lower rank, or conversion to senior civilian positions?

- To what extent should the need to provide developmental experiences for RC G/FOs drive requirements?

A challenging aspect in accomplishing this research was the interrelated aspect of general and flag officer management and accounting. Several terms need to be clarified:

- *Requirements* are positions that the services and Joint Staff designate for a general or flag officer of a particular grade; these positions are usually listed as such in a manpower document.
- *Authorizations* are positions (requirements) that the secretary or chief of the service (for service authorizations) or the Chairman of the Joint Chiefs of Staff (CJCS) (for joint authorizations) has decided to fill with a general or flag officer.
- *Authorized strength* is the maximum number of general or flag officers established in law (for the active component, the reserve component, and each military department); this includes both the basic authorized strength and additional authorized strength in the form of exemptions.
- *General or flag officer inventory* is the number of individuals who can be assigned to fill general or flag officer authorized positions.

What is important to understand is that requirements and authorizations reflect numbers of *positions*, authorized strength is the number of *people* allowed to fill those positions, and inventory is the population of people available to fill the positions—and these numbers may or may not be equal. In fact, requirements are generally greater than authorizations, and authorizations are generally less than or equal to authorized strength. This is because total *requirements* are not constrained by authorized strength limits, whereas filling *authorizations* with a general or flag officer is constrained by the availability of authorized strength.

When we undertook data collection for this study, RC G/FO requirements numbered 634 positions. Authorized strength against these requirements was 595—meaning that the services are authorized to fill only 595 of the 634 stated requirements.[2] Generally, the services authorize RC G/FO positions near the limits of their authorized strength. Each of the services has methods, both formal and informal, to determine which of the requirements to authorize and fill from the inventory of available officers (which we will discuss in more detail in a subsequent chapter). Most of the time, the services will leave a small number of authorized positions unfilled to give personnel managers flexibility to respond to emerging requirements while being able to stay within the limits of authorized strength.

Another challenge was accurately determining which reserve component positions should be included in our review. Congressional language asks for a review of those

[2] These numbers do not include authorizations for RC G/FOs in positions designated for active component personnel in service and joint organizations, which are discussed in more detail in Chapter Two.

RC G/FOs in an *active status*. A reserve component member is considered to be in an active status when he or she is not in the inactive Army National Guard or Air National Guard, on an inactive status list, or in the Retired Reserve. The Reserve Active Status List for each of the military departments lists the officers who are in an active status in the reserve component and not on an active duty list.[3] However, active status applies to *individuals* in the inventory and not to *requirements*. In this review, we included those positions that explicitly require RC G/FOs in the Army National Guard, Air National Guard, Army Reserve, Navy Reserve, Marine Corps Reserve, or Air Force Reserve.

We based our findings and recommendations on information gleaned from a number of sources:

- federal laws that specify basic authorized strength and exemptions for general and flag officers in the active and reserve components
- policy documents guiding the authorizations and management of general and flag officers and positions
- previous studies on the requirements for general and flag officers
- interviews with subject-matter experts and stakeholders in both general and flag officer management offices throughout the Department of Defense (DoD), as well as general and flag officers in both the active and reserve components
- analysis of data collected from the military services on all RC G/FO positions.

Details on specific aspects of our methodology will be discussed in the chapters that follow and relevant appendixes.

Organization of This Report

The remainder of this report describes the results of our review. Chapter Two provides an overview and analysis of the statutory authority for RC G/FOs, including the many exemptions. Chapter Three describes current service approaches to managing RC G/FO requirements, authorizations, and authorized strength and an assessment of whether these approaches are in compliance with existing statutes. Chapter Four presents the approach and results of our analysis of current RC G/FO requirements and concludes with a set of candidate requirements that should be considered for elimination, conversion, or downgrade. Chapter Five explores whether professional development considerations offer guidance to the assessment of RC G/FO requirements. The final chapter discusses the implications of our review and provides recommendations for improving the RC G/FO requirements processes.

---

[3] Title 10 of the United States Code (U.S.C.), Section 14002, directs the Secretary of each military department to maintain a Reserve Active Status List.

CHAPTER TWO

Statutory Authority for General and Flag Officers

Limitations on the total number of general and flag officers in the U.S. military is found in 10 U.S.C., which houses the body of permanent U.S. law relating to the federal military.[1] Title 10 specifies the statutory ceiling for the number of general and flag officers in the active and reserve components (10 U.S.C. 526 and 10 U.S.C. 12004, respectively). Beyond these two sections of law, a complicated raft of other provisions creates exemptions to the ceiling for officers serving in certain types of positions. Over time, Congress has altered the ceiling and created new exemptions. For example, exemptions have been changed to increase the number of officers who can serve in full-time joint positions as a way to meet requirements for joint duty experience for reserve component leaders, and other exemptions have been added for special positions related to a restructured intelligence community following the terrorist attacks of September 11, 2001.

This chapter describes the current statutory authorizations for general and flag officers in both the active and reserve components.[2] It reviews the exemptions in detail, explaining how they can increase the number of RC G/FOs beyond the statutory ceiling. The active duty statutory system is introduced first, because it historically precedes congressional authorization of reserve personnel and because it will serve as a useful point of comparison when the statutory approach to reserve personnel management diverges from that of the active force.

---

1 By "federal military," we distinguish between the armed forces of the United States—including the National Guard of the United States—and the state militias and state National Guard components. 10 U.S.C. 10101 names the Army National Guard of the United States and the Air National Guard of the United States as reserve components of the armed forces. The legislation concerning the state National Guard is primarily housed in Title 32 of the U.S. Code. As a historical matter, the National Guard was primarily and exclusively a state militia force (unless specifically called into federal service) until congressional legislation passed in 1933 created a permanent dual status as recognized state militias and a federal reserve force. See 48 Stat. 153 (1933) (the law sometimes called the National Defense Act of 1933 [Jacobs, 2014, p. 39], the National Defense Act Amendments of 1933 [Kester, 1988, p. 189], or the National Guard Act of 1933 [Rousseau, 1997, p. 11]).

2 Appendix A contains a detailed outline of the key statutory provisions.

Numerical Limits on Active Duty General and Flag Officers

Ultimate responsibility for the size of the officer corps, both active and reserve, derives from the Constitution, which gives Congress the powers to raise and support armies and provide for and maintain navies under Article I, Section 8, and from Congress' general funding, control, and oversight responsibilities. Although the military is under the direction and control of the President, Congress controls the size of the armed forces through explicit limitations on end strength and has done so on many occasions.

Congress regulates the number of general and flag officer positions through statute, with 10 U.S.C. 101(d)(1) defining *active duty* as

> full-time duty in the active military service of the United States. Such term includes full-time training duty, annual training duty, and attendance, while in the active military service, at a school designated as a service school by law or by the Secretary of the military department concerned. Such term does not include full-time National Guard duty.

The number of general and flag officers on active duty is limited by 10 U.S.C. 526. As of 2015, 10 U.S.C. 526(a) sets the authorized strength (number) of general and flag officers on active duty serving in the Army, Air Force, Navy, and Marine Corps at 652, with the allocation across the services as follows:[3]

- Army: 231
- Air Force: 198
- Navy: 162
- Marine Corps: 61.

The allocations to the services are the product of need, historical levels of support, and negotiations among the services, the President, and Congress. Although these allocations would appear to be an absolute limit on the number of active duty general and flag officers, the numbers specified in 10 U.S.C. 526(a) represent only a fraction—albeit a majority—of the active general and flag officers that Congress authorizes. In addition to these allocations, Congress has enacted specific exemptions from the limits on authorized strength, one of which is joint duty requirements. 10 U.S.C. 526(b) sets the authorized strength of general and flag officers on active duty serving in joint duty assignments (referred to as the joint pool) at 310. Adding the joint pool to the service allocations brings the active-duty authorized strength to 962.

---

[3] The authorized limitations *include* the positions of the Chief of Army Reserve (per 10 U.S.C. 3038); Chief of Navy Reserve (per 10 U.S.C. 5143); Commander of Marine Forces Reserve (per 10 U.S.C. 5144); and Chief of Air Force Reserve (per 10 U.S.C. 8038).

Officers falling under the exemptions do not count against the 652 ceiling on active duty general and flag officers, but are additions allowed for specific purposes.[4] Table 2.1 describes the active duty exemptions.

In addition to these exemptions, Congress has authorized a suspension of the limitations on the number of active duty general and flag officers in a time of war or national emergency—as declared by Congress or the President (10 U.S.C. 527).

Table 2.1
Exemptions from Active Duty Authorized Strength for General and Flag Officers

Number of Exempted Officers	Exemption	Statute
310	Officers on joint duty assignment, as designated by the Secretary of Defense	10 U.S.C. 526(b)(1)
15	Positions in the combatant commands for reserve component officers designated by the Chairman of the Joint Chiefs of Staff	10 U.S.C. 526(b)(5)
3	Positions on the Joint Staff for reserve component officers designated by the Chairman of the Joint Chiefs of Staff	10 U.S.C. 526(b)(5)
Unlimited	Reserve component officers on active duty for training or under a call or order specifying a term of less than 180 days	10 U.S.C. 526(c)(1)
Up to 10 percent	Of the authorized strength of RC G/FOs (by armed force, per 10 U.S.C. 12004) on active duty for not more than 365 days	10 U.S.C.526(c)(2)
Up to 5 percent	Of the RC G/FOs, per component, on active duty in excess of 365 days, but not to exceed three years. The limit of 5 positions may be waived with authorization by the Secretary of Defense	10 U.S.C. 526(c)(3)
1	A general or flag officer serving as the Attending Physician to the Congress, if such position is filled by a general or flag officer	10 U.S.C. 526(e)
Unlimited	General and flag officers assigned to temporary joint duty of less than one year, designated by the Secretary of Defense	10 U.S.C. 526(f)(1-2)
Not specified	General and flag officers departing from joint duty assignments, for up to 60 days beginning on the date of departure from the joint duty assignment	10 U.S.C. 526(g)
1	A general or flag officer serving as director or deputy director of the Central Intelligence Agency	10 U.S.C. 528(b)
1	A general or flag officer serving as associate director of military affairs at the Central Intelligence Agency	10 U.S.C. 528(c)
Up to 5	General and flag officers serving in the Office of the Director of National Intelligence, pursuant to agreement between the Secretary of Defense and the Director of National Intelligence	10 U.S.C. 528(d)

[4] There are two types of statutory exemptions. The first simply adds to the 652 limit—the 310 officers that can serve in Secretary-designated joint duty assignments are an addition to the active duty force, for example. The second type of exemption has the effect of assigning general and flag officers to either the authorized strength limit for active duty or reserve component limits. This "accounting" allows for limited numbers of reserve component officers to be used on limited active duty assignment, without depleting the active duty authorized strength.

This set of exemptions from the overall limit on active duty general and flag officers complicates the process of assigning and accounting for active duty officers. Worthy of special recognition is the ability of the active duty force to include a number of RC G/FOs assigned to specific positions that do not count against the active duty end strength limitation, but do count against the RC G/FO limitations (these limitations are described in more detail in the next section). For example, any active component general or flag officer will generally count against the active duty ceiling. However, a reserve general or flag officer on orders specifying an active duty tour for less than 180 days will not be counted against the active duty ceiling. Any general or flag officer on temporary joint duty assignment as designated by the Secretary also would be exempt from the ceiling. Therefore, determining whether a general or flag officer position counts against the active component authorized strength requires a review to assess whether an exemption applies; if no exemption is applicable, the position counts toward the Title 10 limit.

Some of the exemptions are specified as a percentage of the authorized *reserve* officer strength. For example, no more than 10 percent of the reserve officer strength for the Army Reserve can be exempted from the active duty authorized strength when serving between 180 and 365 days. Since the reserve officer strength for the Army (discussed in the next section) is set by law at 207 general officers, this exemption would limit to 20 the maximum number of Army reservists that could be excluded from the active duty limit. Officers beyond 20 would be counted against the limit.

In summary, the number of active duty general and flag officer positions established by law currently stands at 652 officers. These positions may be augmented by additional positions exempted from the limit, as described in various statutes. This end strength relief increases the number of active duty general and flag officer positions under specific conditions. Many of those conditions relate to opportunities for RC G/FOs to serve on active duty or in joint duty positions. Hence, an RC G/FO might serve on active duty, but not count against the active duty limits. Each position has to be checked against the list of possible exemptions to determine whether it counts toward the active duty authorized strength or not.

Numerical Limits on Reserve Component General and Flag Officers

As with active duty officers, RC G/FOs are governed by a set of statutory limits. These include absolute thresholds on the number of general and flag officers and exemptions that specify certain types of positions or conditions that, if met, will not cause specific general and flag officers to be counted against the statutory limit. However, whereas the active duty military is able to add activated reserve officers to its authorized strength (subject to time limitations), the reserve component has no such end strength additions.

10 U.S.C. 12004(a) sets the authorized strength of general and flag officers *in active status* of the reserve components. The Reserve Officer Personnel Management

Act of 1994 established the limit on RC G/FOs at 422—a level that has remained unchanged since that time (Pub. L. 103–337, 1994). The allocation across the services is specified at the following levels:

- Army: 207
- Air Force: 157
- Navy: 48
- Marine Corps: 10.

10 U.S.C. 101(d)(4) defines "active status" as "the status of a member of a reserve component who is not in the inactive Army National Guard or inactive Air National Guard, on an inactive status list, or in the Retired Reserve." The statutory limits for the Army and the Air Force do not specify how general officer positions are to be divided among the Army National Guard and the Army Reserve, or the Air National Guard and the Air Force Reserve. The division of positions across those reserve components is a matter of policy, not of law.

Mirroring the active component general and flag officer ceilings, Congress has enacted exemptions for officers in specific positions that will not count against the limit of 422 set in 10 U.S.C. 12004. Total RC G/FO authorized strength will thus be capped at the sum of the standard statutory limit (set in 10 U.S.C. 12004) plus any general or flag officer personnel in an exempt position. Table 2.2 lists the exemptions.

In addition to these exemptions, Congress has authorized a suspension of the authorized strength limitations for RC G/FOs in a time of war or national emergency, as declared by Congress or the President (10 U.S.C. 12006). There is also provision for four DoD reserve chiefs for the Army, Navy, Air Force, and Marine Corps, which count against the active component authorized strength for general and flag officers (10 U.S.C. 3038, 5143, 5144, 8038).

In summary, the authorized strength for RC G/FOs is established by law in 10 U.S.C. 12004. The basic authorized strength may be augmented by officers filling specific positions as set forth in statutory exemptions. Exemptions are granted for officers serving as either the adjutants general (TAGs) or assistant adjutants general (AAGs), as well as for officers on the National Guard Bureau (NGB). Additionally, the number of RC G/FOs can be up to 20 percent higher than the limit of 422 if the services are using those reservists in joint duty assignments. Finally, the RC G/FO limitations exempt reservists serving on active duty if they would be counted against active duty authorized limits. As with active duty positions, each RC G/FO position has to be checked against the list of possible exemptions to determine whether the officer serving in the position should be counted against the statutory limit.

Table 2.2
Exemptions from Reserve Component Authorized Strength for General and Flag Officers

Number of Exempted Officers	Exemption	Statute
Unspecified	Army and Air Force reserve officers serving as TAGs or AAGs in a state	10 U.S.C. 12004(b)(1)
Unspecified	Army and Air Force reserve officers serving in the NGB	10 U.S.C. 12004(b)(2)
Unspecified	Army and Air Force reserve officers "counted under section 526"	10 U.S.C. 12004(b)(3)
No more than 20 percent of Army or Air Force Reserve general and flag officer positions	Army and Air Force reserve officers serving in joint duty assignments; the number that may be excluded under this exemption may not exceed 20 percent of the authorized limitation for RC G/FO positions for the Army or Air Force, respectively (as specified in 10 U.S.C. 12004)	10 U.S.C. 12004(b)(4)
Unspecified	Navy reserve officers "counted under section 526"	10 U.S.C. 12004(c)(1)(A)
No more than 20 percent of Navy Reserve general and flag officer positions	Navy reserve officers serving in joint duty assignments; the number that may be excluded under this exemption may not exceed 20 percent of the authorized limitation for RC G/FO positions for the Navy (as specified in 10 U.S.C. 12004)	10 U.S.C. 12004(c)(1)(B)
Unspecified	Marine Corps reserve officers "counted under section 526"	10 U.S.C. 12004(d)(1)
No more than 20 percent of Marine Corps Reserve general and flag officer positions	Marine Corps reserve officers serving in joint duty assignments; the number that may be excluded under this exemption may not exceed 20 percent of the authorized limitation for RC G/FO positions for the Marine Corps (as specified in 10 U.S.C. 12004)	10 U.S.C. 12004(d)(2)
Unspecified	Officers departing from joint duty assignment or other non–joint active duty assignment, for up to 60 days beginning on the date of departure from the assignment	10 U.S.C. 12004(f)

Calculating Total Numbers of Reserve Component General and Flag Officers

The exemptions from the authorized limits in 10 U.S.C. 12004 can significantly add to the total number of RC G/FOs. Exemptions for TAGs (54 positions) and AAGs (about 125 at the time of this writing, but the number can vary) in the states and territories collectively add 179 RC G/FOs to the 422 allowed by 10 U.S.C. 12004. Beyond TAGs and AAGs, the officers serving at the NGB (around 25 officers) are also exempt from limits on authorized strength. Furthermore, as previously described, 10 U.S.C. 526 states that the CJCS can designate up to 18 positions to be filled with RC G/FOs (10 U.S.C. 526(b)(5)). These are exempt from the limits set forth in 10 U.S.C. 12004 because of the "counted under section 526" language.[5]

[5] 10 U.S.C. 12004(b)(3)) for Army and Air Force; 10 U.S.C. 12004(c)(1)(A) for Navy; and 10 U.S.C. 12004(d)(1) for Marine Corps.

The remaining exemptions are less specific but they nonetheless add to the authorized strength. For example, the "joint duty assignment" exemptions limit the number of officers who are exempt to no more than 20 percent of the component-authorized strength.[6] Using the limits in 10 U.S.C. 12004 to calculate the maximum number of potential exemptions, there could be up to 41 Army exemptions (207 × 0.2), 31 Air Force exemptions (157 × 0.2), ten Navy exemptions (48 × 0.2), and two Marine Corps exemptions (10 × 0.2). Combined, this has the potential to add up to 84 RC G/FOs above the 10 U.S.C. 12004 limit (and in addition to all of the other exemptions, such as the NGB).

Moreover, the unlimited exemptions for temporary assignments, the exemptions for officers departing joint duty assignments, and the national emergency exemptions are not numerically limited by statute, and can vary because of the fluctuating needs of the country and the timing of personnel departures. For this reason, the total number of RC G/FOs is not static.

The Effect of Exemptions on Reserve Component Staffing, Leadership, and Training Needs

A question we were asked to address is whether this statutory regime of authorized strength with a long list of exemptions is serving reserve component needs. It appears that the intent underlying the special exemptions and the application of the exemptions by the services is meeting a wide variety of staffing, leadership, and training demands of the reserve component.

As mentioned previously, the largest numbers of additive positions are the state leadership positions for TAGs and AAGs. This exemption recognizes the mix of state and federal responsibilities of the National Guard of the United States and the state National Guards. Beyond the federalism dimension, however, the exemptions appear designed to give the services adequate opportunity to share and develop expertise within the RC G/FO corps both on the Joint Staff and in a joint environment. For example, the CJCS's exempt positions are used for developmental jobs in the Joint Staff and in joint and operational environments. Likewise, the exemptions allow the total RC G/FO authorized strength to increase when needed for active duty assignments and in times of national emergency, without running afoul of statutory limits more relevant for peacetime situations.

Nonetheless, the current approach taken in the law and the changing needs of the reserve forces makes an accounting of the total RC G/FO authorized strength a demanding task. Given the flux in the number of officers assigned to various tempo-

---

[6] 10 U.S.C. 12004(b)(4) for Army and Air Force; 10 U.S.C. 12004(c)(1)(B) for Navy; and 10 U.S.C. 12004(d)(2) for Marine Corps.

rary duties, the number of authorized RC G/FOs is constantly changing. The exemptions, as they are written, give the reserves a great deal of needed flexibility. In Chapter Three, we discuss the practical implications of managing these exemptions and whether changes to the exemption framework might be helpful.

CHAPTER THREE

Managing Requirements, Authorizations, and Authorized Strength

To better understand how the services manage their general and flag officers, we spoke with a number of subject-matter experts across DoD—from the active and reserve components of all four services, DoD and service organizations that manage general and flag officer positions, and other relevant organizations—who provided information on how requirements for RC G/FO positions are generated, how authorized strengths are managed per the requisite regulatory statutes, and the service's philosophies toward managing and developing general and flag officers.[1] From this information, RAND researchers were able to draw a number of conclusions about the status of RC G/FO requirements determinations, individual development, and overall management philosophies, which are discussed in this chapter.

From Requirements to Authorized Strength

The service and Joint Staff processes used to generate and validate requirements, authorize positions, and allocate authorized strength for RC G/FOs vary in detail but perform similar functions overall.

- Organizations in the active component, the reserve component, the Joint Staff, the combatant commands, or defense agencies—individually or in coordination—develop job descriptions for positions believed to require an RC G/FO, which are submitted to the RC G/FO management offices of the services and the Joint Staff, who validate those requests.
- The RC G/FO management offices report that they assess a number of characteristics in validating requests for new positions: the nature of the position's duties and the magnitude of its responsibilities, the span of control and scope of the resources managed, the significance of actions and decisions required by

[1] Appendix B lists the organizational affiliation of the subject-matter experts who participated in interviews and the complete interview protocol.

the position, the importance of mission accomplishment to national security and other national interests, the level of subordinates within the organization, and the grade of those in similar positions. However, no formal classification standards or guidelines are evident in the process and none of these classification considerations are tied to specific measurements. These characteristics are considered generally and very broadly to provide significant latitude for service leadership judgment when making the decision to create (or validate) a general or flag officer position. All of the general and flag officer management offices report that the single most important consideration in validating a general or flag officer position is the military judgment of service and joint senior leaders.

- The Secretary of Defense, Chairman of the Joint Chiefs of Staff, Service Secretaries, Service Chiefs, and Chiefs of the Reserve and NGB ultimately determine whether a position should be authorized and filled by an RC G/FO who will count against the overall authorized strength allowed by Congress. Each time a decision is made to fill a new authorization, a decision must also be made to leave another position vacant in order to stay within authorized strength limits—a process that is often ad hoc.[2]

The majority of the reserve components do not report any periodic revalidation of all authorized positions. The one exception is the Air Force Reserve, which uses a transparent and well-documented system for validating requirements and prioritizing which of their general officer positions should be filled within the limits of authorized strength (Air Force Reserve 2014 General Officer Position Study Group [GOPSG], 2014). The Air Force Reserve GOPSG meets every two years to consider new general officer position requests and revalidate existing general officer authorizations, given overall stated requirements. Its goal is to maximize Air Force Reserve support to Air Force priorities. The study group is made up of Air Force Reserve major generals and reports through the Chief of the Air Force Reserve to the Air Force Chief of Staff for approval.

The study group uses four criteria, as noted in Table 3.1, to evaluate all general officer positions, and it determines weights for the four criteria based on mission requirements.[3] Once the weights are set, the positions are rank-ordered based on their score using the weighted criteria. Based on these rankings, the study group allocates authorizations, leaving some positions unfilled, to be considered by a future GOPSG for an authorization. The criteria weights from the 2014 GOPSG are shown in Table 3.1.

There are slight differences in the processes used to establish requirements and ensure statutory compliance with authorized strength within the services, the National Guard, and joint and component command staffs, as described in Table 3.2.

---

[2] We are not referring here to a temporary vacancy in a position due to the reassignment of the incumbent before a successor is available. Rather, there is a set of positions that the services and the Joint Staff see as current and funded. We are referring to the ad hoc prioritization of the positions they intend to fill.

[3] The study group uses paired comparisons to set the weights.

Table 3.1
Air Force Reserve General Officer Position Study Group Weighted Criteria

Criteria	Criteria Description	Weight (%)
Support to the warfighter	Support to functions providing direct, operational-level support to the warfighter Significant Title 10 responsibility to organize, train, and equip forces in accordance with U.S. Air Force priorities Other direct warfighter support through enablers such as acquisition, weapon system sustainment, and readiness.	39.17
Advocacy/ policy influence	Support to organizations that: • provide the ability to advocate reserve component interests in strategy, policy, planning, programming, and budgeting • promote advocacy at the strategic, operational, or tactical level • provide an Air Force Reserve counterpart to an Air National Guard general officer of a similar grade Frequency/likelihood of opportunities for filling in for the principal	35.29
Senior Air Force Reserve advisor role	Extent and impact of official duties as the organization's senior Air Force Reserve advisor	7.35
Major command/ functional area ranking	The active component's prioritization	17.65

SOURCE: U.S. Air Force Reserve General Officer Management Office.

The Joint Staff General/Flag Officer Matters Office manages all general and flag officer positions assigned to the Joint Staff, Office of the Secretary of Defense (OSD), combatant commands, and defense agencies. Of these positions, 310 are called the *joint pool* because, within certain constraints, they can be held by general and flag officers from any service and either the active or reserve component. The Secretary of Defense has delegated to the CJCS the authority to manage 294 of the 310 positions. A position can be added to the joint pool at the request of a combatant command or DoD agency with CJCS approval. Current CJCS guidance is silent on the criteria to consider when submitting a request for a new general or flag officer position (CJCS, 2010). However, since the Secretary of Defense has instituted a "no general or flag officer growth policy," a compensatory offset will be required if the joint pool lists a full complement of 294 positions, and it is typically expected that the offset will come from the requesting organization. As a result, there is an ad hoc process for prioritizing positions that occurs whenever a new requirement is identified.

The Joint Staff General/Flag Officer Matters Office is also responsible for the Chairman's Reserve Positions—15 on combatant command staffs and three on the Joint Staff. These positions are full-time, nominative reserve component brigadier general and major general authorizations under control of the CJCS for distribution (CJCS, 2010). Again, current CJCS guidance does not provide criteria for choosing when new joint positions should be validated and authorized.

Table 3.2
Details on Requirements and Authorization Processes

Organization	Overall Requirements Processes	Authorization Distribution
Army and Air National Guard	State/territory-dependent definition of number, organization, grade, and competencies required for state TAGs and AAGs No written documentation on requirements processes at the NGB–level were presented For general officer-of-the-line requirements in active component and other reserve component organizations, the NGB coordinates on the military departments' requirements processes The chief of the NGB approves NGB headquarters requirements at the time the need for a new position emerges	Operates at limits of basic authorized strength; operates below the limits of exemptions Statutes do not specify the proportion of Army authorizations that should be allocated between the Army National Guard and the Army Reserve, or the proportion of Air Force authorizations that should be allocated between the Air National Guard and the Air Force Reserve; a historical division of the authorization is being used; no mechanism or process was presented during our review for determining the appropriate ratio Using wartime and national emergency exemptions for additional authorizations (as of February 2015, three Army National Guard general officers are on full-time duty exempt from active component strength ceilings)
Army Reserve	No formal, documented reserve component general officer requirements processes Any agency that needs a position for a reserve component general officer submits a request to the Office of the Chief of the Army Reserve and the Chief of Staff of the Army In the case of predefined unit configurations, Army G-3/5/7 (Operations and Plans) validates general and flag officer position requirements using standard, well-documented manpower processes No documentation for the types of positions/responsibilities that call for a general officer position Decision on whether a general officer is required for a position is made when new requirement emerges	Operates at limits of basic authorized strength; operates near limits of exemptions Using wartime and national emergency exemptions for additional authorizations (as of February 2015, two Army Reserve general officers are on full-time duty exempt from active component strength ceilings)

Table 3.2—Continued

Organization	Overall Requirements Processes	Authorization Distribution
Navy Reserve	No documentation for the types of positions and responsibilities that call for reserve component flag officers; no written documentation on flag officer requirements process specifically Requirements established by the active component (Chief of Naval Operations) and filled by reserve component (Commander, Navy Reserve Force) Decision on whether flag officer is required for a position is made when new requirement emerges	Operates below the limits of basic authorized strength; operates below limits of exemptions Strictly follows statutory limits and purposely limits the number of authorized flag officers below the statutory limit to ensure management flexibility for more flag officers
Marine Corps Reserve	Requirements are established by the active component (Commandant of the Marine Corps) and filled by Commander, Marine Corps Forces Reserve New reserve component general officer positions are created to support active component changes in mission or structure No specific documentation on what types of positions reserve component general officers should fill Marine Corps senior leaders periodically review all positions and determine if there is a need to support new and emerging general officer requirements	Operates below limits of the basic authorized strength; operates below the limits of exemptions Accounting at the individual level is simplified by the small number of general officers authorized (approximately ten) More requirements than available authorizations are purposely established and maintained; senior leaders decide which of the positions to allocate the authorizations to depending on the needs of the active component
Air Force Reserve	Air Force Reserve leaders produce a biannual prioritization of general officer requirements. Process includes criteria for the responsibilities that determine whether a position requires a general officer Quarterly executive review (Chief of the Air Force Reserve) of future requirements	Operates at the limits of basic authorized strength; operates below the limits of exemptions; purposefully does not use full allowance of authorized strength to allow for management flexibility

Officials for the general and flag officer management offices in each of the military departments report that slightly more requirements are designated for general and flag officers than authorized strength limitations will allow them to fill. They all report that they are able to fill requirements to a sufficient level to meet mission needs and that critical requirements are not going unfilled due to insufficient authorized strength. The day-to-day attention of the management office staffs ensures that mission needs are met by identifying appropriate requirements, creating well-written position descriptions, assigning qualified officers to fill open positions, and managing the available authorized strength.

Though the requirements processes generally work, they do have the potential to result in inconsistencies across services because of differences in individual service philosophies on the appropriate roles for RC G/FOs. These differences can, for instance, result in positions with similar responsibilities being filled by individuals of very different rank. Within each service, however, there is a degree of consistency.

Compliance with General and Flag Officer Authorized Strength

The general and flag officer management offices expend considerable time and attention ensuring that they are adhering to the complex set of statutory limitations to general and flag officer authorized strength. Each management office has its own method for tracking assignments. We audited these tracking systems against our independent assessment of the statutory authorized strength and exemptions (described in Chapter Two). Although the services use different terminology and accounting approaches, we found that these systems accurately, appropriately, and consistently ensure compliance with statutory limits on the number of RC G/FOs. In reviewing these tracking systems, and in discussions with the management offices, we observed that each of the reserve components is operating at or near its limit for the number of RC G/FOs, given the limitations imposed by the congressional authorized strengths.

Appropriateness of Exemptions to Authorized Strength

RC G/FO management offices report that the roles of RC G/FOs are more varied than those of general and flag officers in the active component. For example, RC G/FOs are called upon to

- serve as reserve component representatives and integrators on service and joint staffs
- serve as mobilization assistants
- lead reserve commands and organizations

- fill active component skill gaps
- provide ready assets for war and national emergency
- lead at the state level (National Guard, Title 32)
- lead and staff the NGB
- fill needs for new and emerging skills
- participate in developmental experiences.

The RC G/FO management offices report that the current system of statutory authorizations and exemptions, as discussed in Chapter Two, generally allows them to fill these roles. The exemptions, in particular, are targeted to provide the additional authorizations—both in type and number—needed to meet some of these requirements and provide essential flexibility to the services. That said, some managers thought that a simpler system—one that combined some exemptions and the basic authorized strength into a single pool—would be easier for the services to understand and to manage. They suggested it could also mitigate the perception that there are "too many" RC G/FOs. Figure 3.1 offers a notional representation of one method for simplifying exemption statutes.

In considering this proposal, RAND researchers would not recommend pooling together TAGs, AAGs, and NGB authorization exemptions. These exemptions are used for very specific purposes: (1) to authorize general officers under the control of the states and territories that hire and retain TAGs and AAGs and (2) to staff the National

Figure 3.1
Notional Consolidation of RC G/FO Basic Authorizations and Exemptions

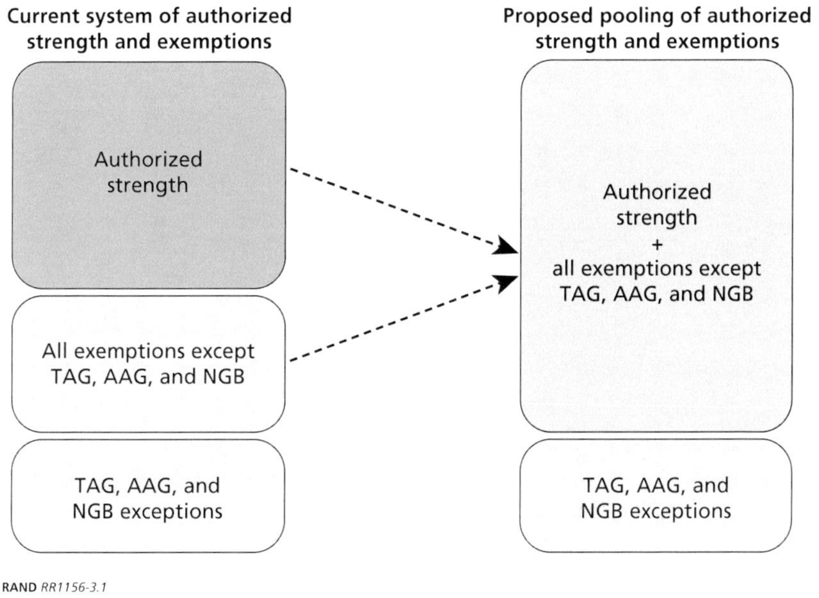

Guard under the control of the Chief of the NGB. As a result, they should not be part of a broader pooling scheme. Aside from this, there are pros and cons for pooling the remaining positions.

On the positive side, consolidating exemptions with basic authorized strength would

- provide a transparent authorized topline
- make it easier to track and report authorized strength
- fix the total authorized strength number, rather than allow it to vary depending on where RC G/FOs are assigned
- allow the services more latitude in deciding how the total number of authorizations should be distributed among the varying roles for RC G/FOs
- obviate the need for the legal interpretations that have been necessary in reconciling the active and reserve component exemptions (10 U.S.C. 526 and 10 U.S.C. 12004).

On the negative side, consolidating exemptions with basic authorized strength would

- change an already implemented and generally understood process.
- eliminate specific categories of exemptions that could have unintended consequences resulting from interactions with other sections of law. (For example, consolidating exemptions with basic authorized strength might limit the number of RC G/FOs serving in the joint pool.)
- eliminate exemptions that are designed to encourage the use of RC G/FOs in certain types of jobs, such as joint jobs. (With a single pool, the purposes behind specific exemptions would no longer be evident and RC G/FOs might no longer be used in those roles.)
- be inconsistent with active component authorizations and exemptions.

The process associated with any statutory change is inherently uncertain. Congressional action taken to pool basic authorized strength and exemptions could result in a lower total authorized strength for RC G/FOs. There is also the potential for losing the ability to explicitly designate individuals to fill particular roles (which are now provided for by exemptions). As a result, we cannot find sufficient reason to support changing the current system of basic authorized strength and exemptions.

Additional Issues Potentially Affecting RC G/FO Requirements, Authorizations, and Authorized Strength

Converting Positions to Senior Executive Service

One suggested avenue for controlling growth in the number of RC G/FOs is to convert some of the positions to senior civilian positions and potentially assign members of the Senior Executive Service (SES) to meet the requirements. The stakeholders we talked to did not see an opportunity for converting additional RC G/FO positions to SES. They told us that in many of these positions, the seniority carried by a military general or flag officer was necessary for effective interactions with other senior military officials and stakeholders. For positions with specific military command requirements, an SES member would not be appropriate. In addition, many requirements are part-time positions or may require deployment. The Civilian Senior Executive Management Division agreed that conversion for part-time positions would be problematic. However, in Chapter Four, we do analyze the possibility of converting RC G/FO positions to SES further, using data on RC G/FO positions gathered from the services and joint organizations.

Opportunities for Officer Development

Individuals assigned to general and flag officer positions must have certain competencies that can be obtained from training and experience. Both the active and reserve component general and flag officer management offices expressed the need for positions that can provide developmental opportunities for RC G/FOs so that an appropriately developed pool of officers is available for particular functional requirements and for increased levels of responsibility in the future.

The Air Force Reserve, the Navy Reserve, the Army Reserve, and the Marine Corps Reserve all report that they have a sufficient number of authorizations for adequate development of the current pool of general and flag officers. The National Guard reported that it lacks opportunities to develop officers adequately for more senior positions, including the Chief of the NGB, and would like additional authorizations for that purpose. To ensure we were appropriately considering professional development needs, we looked at this topic in depth, including consulting with currently serving RC G/FOs. A description of this review and our conclusions are in Chapter Five.

Relating General and Flag Officer Authorized Strength to Overall End Strength

Debate about the appropriate authorized strength for general and flag officers—in the active and reserve components—has often included comparisons between general and flag officer authorized strength to total service or component end strength. Figure 3.2 depicts the total number of personnel in the active component and the selected reserve

Figure 3.2
Active Component and Selected Reserve Personnel, FYs 1980–2014

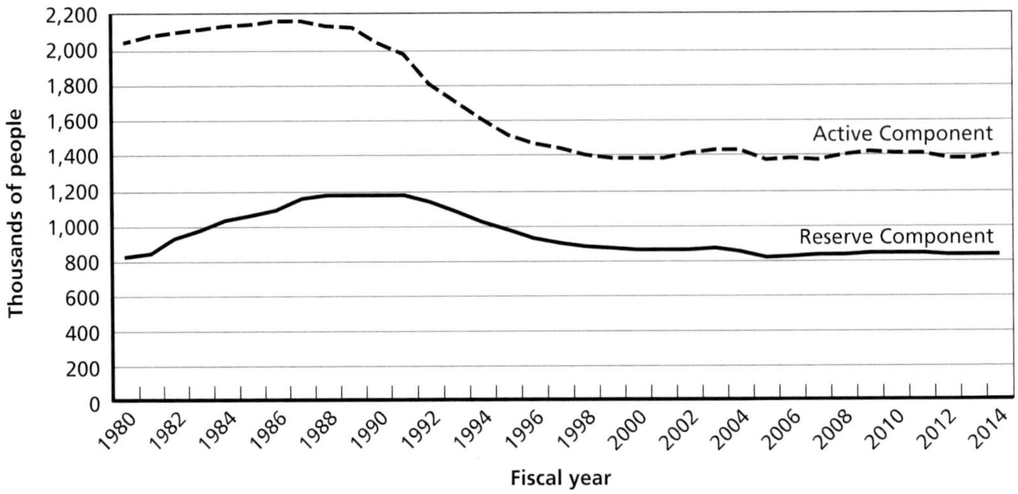

SOURCES: Defense Manpower Requirements Reports (2000–2015) and Acting Assistant Secretary of Defense for Force Management Policy (2001).
RAND *RR1156-3.2*

Figure 3.3
Ratio of RC G/FO Authorized Strength to Selected Reserve Personnel, FYs 1980–2014

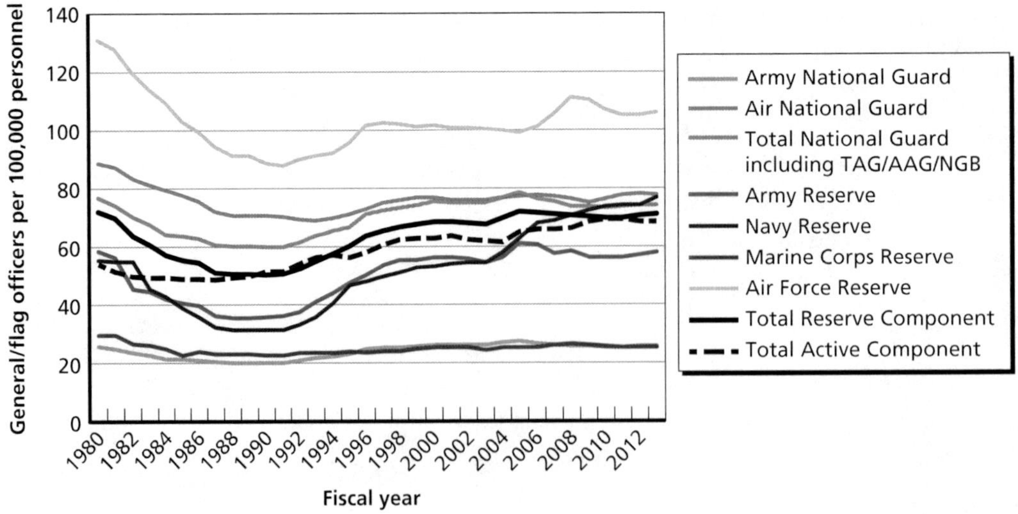

SOURCES: Defense Manpower Requirements Reports (2000–2015) and Acting Assistant Secretary of Defense for Force Management Policy (2001).
RAND *RR1156-3.3*

from fiscal years (FYs) 1980–2014.[4] Following the end of the Cold War, both the active and reserve components experienced significant reductions in end strength—although cuts in reserve end strength were less severe than those for the active component. During this drawdown, the basic statutory authorization for RC G/FOs was unchanged. Consequently, the ratio of authorized RC G/FOs to total selected reserve personnel increased for most of the reserve components beginning in the early 1990s, as shown in Figure 3.3.

The Air Force Reserve has the highest ratio, at 105 general officers for every 100,000 selected reserve personnel in FY2014. The Navy Reserve experienced the greatest increase in this ratio, rising from 31 reserve component admirals for 100,000 selected reserve sailors in FY1991 to 74 admirals for 100,000 sailors in FY2014. Exceptions are the Army National Guard and Marine Corps Reserve, which have remained relatively steady for more than three decades.

The Army National Guard and Air National Guard ratios in Figure 3.3 do not include TAGs, AAGs, and generals on the NGB staff, which are exempted from statutory limitations. Instead, these positions are included in the Total National Guard ratio, which is more appropriate given the role that National Guard general and flag officers play in the leadership of state and territorial personnel and missions. This ratio has increased less drastically than the ratio for the Army and Air National Guard, rising from 60 general officers for 100,000 selected reserve in FY1990 to 78 general officers per 100,000 in FY2005, with slight decreases since then to 74 in FY2014.

Force size alone is not an accurate determinant of general and flag officer requirements, so authorized strength should not automatically fluctuate with changes in reserve component end strength. Much of the need for military executive leadership derives from management responsibilities that exist whether the forces are large or small. For example, training commands need executive leadership regardless of how many forces are being trained. In addition, senior officers today have responsibilities not only in combat but also in technological and political spheres that demand the deep experience and political and operational sophistication characteristic of general officers.

Another reason the overall ratio between the size of the force and the number of general and flag officers has grown over the past few decades is the advancement in technical warfare that began during the Cold War. The services calculate the required supervision ratios of general and flag officers against the increasing destructive capability delivered by one aircraft, ship, or combat-armed individual, and accepting lower ratios may introduce unacceptable risk.

---

[4] The reserve component comprises the active reserve, the inactive reserve, and the retired reserve. The selected reserve is that portion of the active reserve currently participating in training in those units and positions prepared to mobilize, and includes full-time support personnel.

Evaluating authorized strength for RC G/FOs based on the total component end strength is particularly problematic. Valid roles for RC G/FOs do not always put them in command of large reserve component units, either for training purposes or upon mobilization. Reserve component units and individuals can integrate with or backfill active component members, augment headquarters staffs to provide expertise, facilitate training or mobilization processes in areas such as logistics or transportation, and perform state missions. RC G/FOs fill in for shortfalls in active component general and flag officer positions, either with particular skills or in times of war or national emergency. In addition to serving in positions within each reserve component, as TAGs and AAGs, and at the NGB, RC G/FOs also serve on the Joint Staff and at combatant commands. And in any of these positions, RC G/FOs may supervise only a small number of personnel.

This discussion considers only the basic authorized strength for RC G/FOs, totaling 422 across the military departments. The authorized strength provided for in the exemptions varies over time, by service and component and, therefore, requires examination of the actual number of RC G/FOs over time, as shown in Figure 3.4. Between FY 1996 and 2014, the number of RC G/FOs increased by 18 percent to a total of 647. This increase reflected the changing leadership role of the reserve component in the total force, especially after the 9/11 attacks, as RC G/FOs were more frequently assigned to positions traditionally filled by active component general and flag officers

Figure 3.4
RC G/FO Inventory, FYs 1996–2014

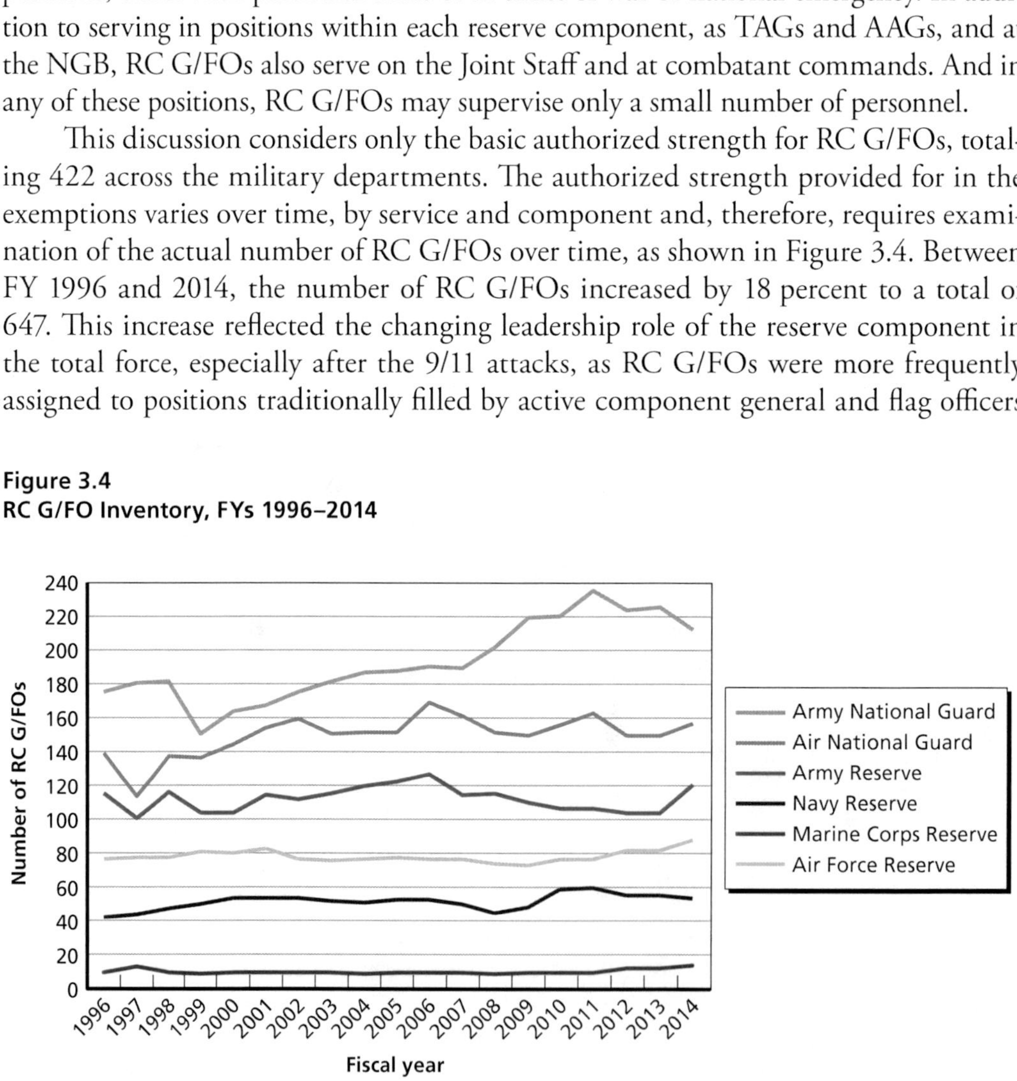

SOURCES: Defense Manpower Requirements Reports (2000–2015) and Acting Assistant Secretary of Defense for Force Management Policy (2001).
NOTE: The inventories for the Army National Guard and the Air National Guard include TAGs and AAGs. Not all of these TAGs and AAGs were eligible to serve in the Title 10 positions designated as General Officer of the Line.
RAND *RR1156-3.4*

and as their presence has become increasingly valued on joint and combatant command staffs. Statutory exemptions actually encourage the participation of RC G/FOs in roles where they are not leading large numbers of personnel; limiting exemptions based on end strength might mean that these key roles for RC G/FOs would be unsupported.

Statutory Exemptions in Time of War or National Emergency

Two sections of law provide for the use of RC G/FOs in times of national emergency or war. 10 U.S.C. 12006 gives the President the authority to suspend RC G/FO strength limitations (422 RC G/FOs) when additional officers are needed. But these are part-time positions, and it is difficult to envision a situation in which a national emergency would require additional part-time officers. A more likely circumstance would be a need for additional full-time RC G/FOs during a time of war—for mobilization, to support mobilization, to lead newly formed units or organizations, or to backfill active component general and flag officers. Except for certain exemptions, RC G/FOs are counted toward active component strength limitations when serving full time (10 U.S.C. 525). Therefore, to increase the number of full-time RC G/FOs during times of war or national emergency, active component strength and grade limitations would need to be suspended, as is authorized in 10 U.S.C. 527 and 10 U.S.C. 525.

The current Presidential Executive Order (White House, 2001) delegates approval for 10 U.S.C. 527 exemptions to the Secretary of Defense. The Office of the Under Secretary of Defense for Personnel and Readiness (OUSD[P&R]) has issued additional guidance that "activations in excess of general and flag officer grade and strength ceilings should be kept to the absolute minimum required to meet mission needs" (OUSD[P&R], 2003). There are three types of requests for wartime or national emergency exceptions to general and flag officer strength ceilings:

- unit commanders—as part of a unit activation
- directed mission—in cases where the military department has been directed to assume a mission
- individual skills—as a request for an individual with specific skills to meet a predetermined requirement.

Figure 3.5 shows the use of 10 U.S.C. 527 for RC G/FOs for available reporting periods from 2006 to 2014. The Army has made most use of this statutory relief from active component strength limitations. The Army reports that since 9/11, a maximum of 47 full-time reserve component general officers were exempted from active component strength ceilings at any point in time.[5] As of February 2015, two Army Reserve and three Army National Guard general officers serve on full-time duty exempt from active component strength ceilings. Although the military departments have found it

---

[5] This maximum occurred prior to the period shown in Figure 3.5.

Figure 3.5
Number of RC G/FOs on Full-Time Duty Under 10 U.S.C. 527, 2006–2014

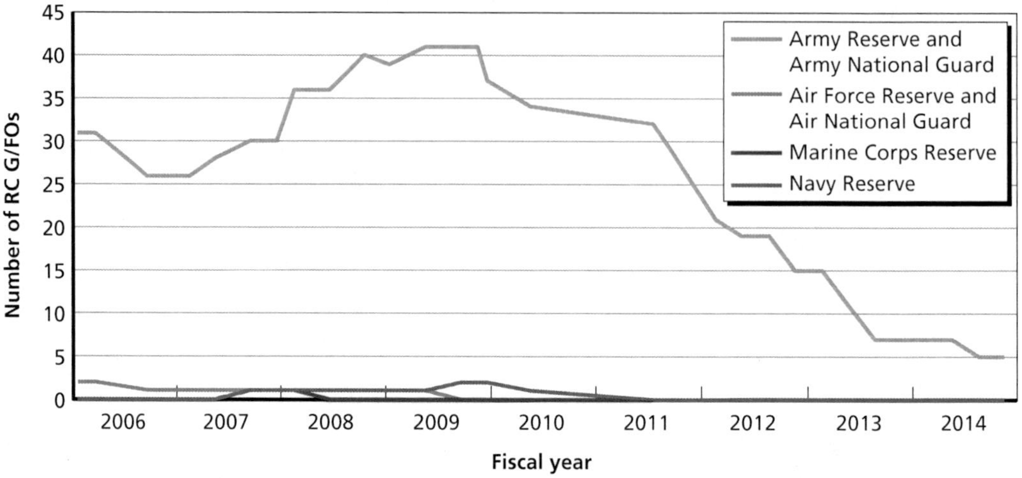

SOURCE: Unpublished data provided by Office of the Under Secretary of Defense for Personnel and Readiness OUSD (P&R) Officer and Enlisted Personnel Management (OEPM).
RAND *RR1156-3.5*

necessary to bring RC G/FOs on full-time active duty to support their needs for senior leaders, none of the military departments or components reported that they needed additional numbers of RC G/FOs to support these full-time needs. All say that there is a sufficient pool of part-time unit commanders and especially mobilization assistants available to activate. Thus, while statutory exemptions are useful to the service in time of war or national emergency, these exemptions, like component end strength, should not drive RC G/FO authorized strength.

Findings and Recommendations

All previous general and flag officer requirements studies we reviewed emphasize the difficulties in determining how positions requiring general and flag officers are designated or how many officers are needed. We recognize that military senior leaders need to have the latitude to make these decisions consistent with how the services organize, train, equip, and employ forces. Nevertheless, some general guidelines that apply to all the services and common systems used by all the services could be useful in achieving greater consistency in how these positions are managed departmentwide. Toward that end, we offer the following recommendations:

- **Develop guidelines for evaluating and validating reserve component general and flag officer requirements.** We recommend that the OUSD(P&R) develop and release guidelines for evaluating and validating positions for RC G/FOs based on the roles and responsibilities for these officers. Criteria should encourage consistency across service and joint positions while providing adequate flexibility to meet individual mission needs of the services, the Joint Chiefs of Staff, OSD, and the combatant commands. A formal process for creating or validating RC G/FO positions, such as the one being used by the Air Force Reserve, would provide greater visibility into and justification for these positions in the future. The criteria described in Chapter Four can provide the basis for developing such guidelines.

- **Institute methods for DoD-wide tracking of reserve component general and flag officer authorized strength.** Our evaluation determined that the services and Joint Staff have appropriate and accurate methods for ensuring compliance with statutory limitations on the number of RC G/FOs. Still, their methods of accounting are not consistent and it took significant time and coordination to ensure that they were all interpreting and adhering to statutory limitations. We recommend that OUSD(P&R) institute a system for tracking the RC G/FOs provided for under the basic authorization (422 total under 10 U.S.C. 12004) and for the individual categories of exemption using a consistent taxonomy across the reserve components. With a common tracking system used across DoD, OUSD(P&R) can more easily ensure that the number of RC G/FOs does not exceed statutory limits and can more accurately track the use of exemptions over time. A common tracking system should foster increased collaboration between the active and reserve component general and flag officer management offices to share best practices in making best use of authorizations and exemptions.
- **Maintain the current system of basic authorizations plus individual exemptions.** While the general and flag officer management offices reported that the structure of statutory authorizations and exemptions is complex and difficult to manage, we did not find sufficient evidence to support changes to statute. The current categories of exemptions provide for necessary full-time, developmental, and leadership roles for RC G/FOs and give the services flexibility to use reserve component officers in the way that best meets evolving mission needs.
- **Resist attempts to link reserve component general and flag officer limits to end strength.** We reviewed two factors that have the potential to affect the authorized strength for RC G/FOs—end strength numbers and the use of RC G/FOs in times of war or national emergency. We found that neither factor drives the need for increased authorized strength for RC G/FOs. As such, limiting RC G/FO authorized strength based on end strength would be counterproductive.

CHAPTER FOUR

Review of General and Flag Officer Positions

The previous two chapters described the statutory authority for RC G/FOs and how the military services manage and staff positions designated for these senior officers. Our analysis shows that the services are operating within the limits prescribed in the law, including the many exemptions to the basic authorized strength. The next step in our assessment of RC G/FO requirements was to evaluate whether all of the stated service requirements are needed or whether some positions could be eliminated, downgraded, or converted to senior civilian positions.

To address this issue, RAND researchers worked in coordination with OASD/RA, the services, and the Joint Staff to collect data on existing reserve component positions requiring a general or flag officer (as of February 2015). That is, any position that (a) requires a general or flag officer and (b) must be filled by a person coming from one of the reserve components. Using these data, we reviewed the positions against a variety of factors that historically have been associated with the need for a general or flag officer.

This chapter presents a brief overview of the methodology used in RAND's assessment, a description of the data, the results of our analysis, and recommendations.

Data Collection Methodology and Protocol

RAND researchers reviewed several previous general and flag officer studies to inform the data collection methodology and protocol. The primary studies included the following:

- Hay Group, *Study of General/Flag Officer Requirements and Distributions in the Department of Defense*, Washington, D.C., 1988.
- Kapos Associates, *Analysis of U.S. Marine Corps Reserve General Officer Billet Requirements*, Arlington, Va., August 31, 1996.
- GAO, *General and Flag Officers: DoD's Draft Study Needs Adjustments*, April 8, 1997a.

- 2003 LMI study of general and flag officers, cited in GAO, *Military Personnel: General and Flag Officer Requirements Are Unclear Based on DOD's 2003 Report to Congress*, GAO-04-488, April 2004.
- Internal DoD study on general and flag officer efficiency, unpublished, 2010.
- U.S. Senate, *General and Flag Officer Requirements: Hearing Before the Subcommittee on Personnel of the Committee on Armed Services*, Senate Hearing 112-258, Washington, D.C., September 14, 2011.

We also reviewed two other GAO reports to assess methodologies and identify weaknesses or deficiencies:

- *Military Officers: Assessment of the 1988 Defense Officer Requirements Study*, GAO/NSIAD-88-146, April 1988.
- *Military Personnel: General and Flag Officer Requirements Are Unclear Based on DOD's 2003 Report to Congress*, GAO-04-488, April 2004.

All of these studies used formalized "job evaluation methodologies."[1] In its report on the 2003 DoD study, the GAO highlighted the key limitation of this approach: that key elements of the process are predominantly based on subjective judgment (GAO, 2004, p. 13) and could lead to inconsistent scoring and results. The selection of measurement factors, the assessment of many (but not all) of those factors, and the weighting scheme used to combine factors into an overall score all depend on subjective judgment; however, GAO did not offer a specific method that they would qualify as objective. GAO also highlighted other limitations of job evaluation methodologies: They account for an organization at a fixed point in time and may fail to capture emerging needs, and they often do not account for dual-hatted positions. The other limitation pointed out by GAO was that all previous studies take the current force structure as a given.

The 1988 study by the Hay Group was designed to determine and validate general and flag officer requirements and distributions. It employed the "Hay Method of Job Evaluation," which is a formal procedure for ranking a set of jobs by their worth to an organization—a methodology that many subsequent studies also utilized. The Hay method focused on job context and content. The primary factors used for job measurement were know-how, problem solving, and accountability. Of note, the Hay Group also conducted a small sample of similar evaluations of SES and O-6 positions for comparison. One deficiency of the Hay Group study was that it did not include mea-

---

[1] This approach is believed to have its origin in 16 specific factors identified by the Bolte Commission in the late 1950s, although these factors do not appear in the commission's reports. The earliest known listing of the 16 factors appears in two service regulations: Marine Corps Order 5311.4, September 30, 1986, and OPNAV Instruction 5420.87A, February 12, 1981 (GAO, 2004).

surements of external components of the job, such as parity with foreign or domestic counterparts or ability to enter into agreements with foreign governments.

Kapos, in its 1996 review of Marine Corps general officer positions, used a methodology involving a multistep job evaluation process that included 21 attributes (listed in Appendix C) as criteria to determine if a position merited a general officer.[2] Each position was evaluated to determine the presence or absence of those attributes, which were divided into four broad groups: nature of position, magnitude of responsibilities, significance of duties, and special qualifications. The methodology also divided the organization into five sectors: operations, combat development, material, service headquarters, and out-of-service. Using the criteria, a panel then determined which positions exhibited which attributes. Finally, the study group developed a standard of how many attributes a position in each of the organizational categories would need to have in order to constitute a general officer requirement.

Both the Kapos and Hay Group methodologies were used in the 1997 DoD review of general and flag officer positions. The services used contractors to independently conduct reviews of their general officer and flag officer requirements. The Army, Navy, and Marine Corps all used variations of the Kapos methodology and the Air Force used a version of the Hay methodology. Because of the inherently subjective nature of scoring positions, GAO found that differing methodologies made cross-service comparisons difficult, even when comparing similar positions from one service with another, and indicated that a single methodology, consistently applied, would have been a better approach (GAO, 2004, p. 21). However, GAO also criticized the review for double-counting some requirements as a result of both the services and the Joint Staff providing input and for failing to identify candidates for military-to-civilian conversion.

The 2003 LMI review built on the Hay Group job evaluation methodology and addressed some of its criticisms. The review validated positions based on 16 factors (see Appendix D), then rank-ordered all of the positions. For the most part, GAO commended the LMI review, suggesting that it improved on some of the criticisms of the 1997 general and flag officer study. The most marked improvement was the use of a single methodology and a common set of factors to evaluate positions across the services and joint community. The report did not assess positions that could be converted from military to civilian positions, but it did recognize the need to identify those positions. In addition, because each service compiled its own data, the factors used by the LMI study were open to different interpretations that caused inconsistencies when comparing similar general and flag officer positions.

The 2010 DoD General and Flag Officer Efficiencies Study Group, mandated by the Secretary of Defense, conducted a baseline review of all active general and flag offi-

---

[2] According to GAO (1997), the Kapos methodology includes up to 25 criteria, but only 21 items were used in the 1996 study.

cer positions. It differed from other studies that RAND researchers reviewed, in that it was specifically focused on identifying at least 50 positions for elimination rather than determining how many positions were required. As part of the study, the services were asked to provide an evaluation of all general and flag officer positions using the following criteria:

- tiered prioritization from 1–4 (1–must have, 2–need to have, 3–good to have, 4–nice to have)
- line of operation (operations, headquarters, service support)
- category (military operation-direct action, military support, military presence, military experience).

The DoD methodology built on the job evaluation methodologies of previous studies but also recognized the many limitations of that approach. To bolster its results, the committee established a panel to provide insight as to organizational structure, reasons for creation of certain organizations, and reasons that general officers are required at the ranks they are designated.

After reviewing the relevant literature, RAND researchers decided to use the 2003 LMI study as the basis for their data collection protocol because this study improved on many areas in which previous studies were criticized and received generally favorable reviews from GAO. To address the weaknesses of the LMI study documented by GAO and others, RAND researchers reworded and added questions to complement and supplement the 16 factors used by LMI. The full RAND data collection protocol is provided in Appendix E; Appendix F contains details on the development of the RAND data collection protocol, the data collection process, and how RAND researchers coded the data.

Analysis of Current Reserve Component General and Flag Officer Positions

In examining current general and flag officer positions, our intent was not to conduct a zero-based, detailed review and assessment of each position that employs an RC G/FO. Rather our aim was to identify subsets of positions that might offer opportunities for elimination, downgrading, or conversion subject to further service and Joint Staff review. To do this, we "inverted" the approach taken by prior studies. Prior studies that reviewed general and flag officer positions were all based on a similar underlying principle: If a position has *enough* factors associated with a general or flag officer, then the requirement is justified. In our inverted methodology, a position that *doesn't have any* of the key factors associated with a general or flag officer (factors are described later in this chapter) becomes a candidate for detailed review by the services to determine if it in fact should be eliminated, downgraded, or converted. Notwithstanding prior attempts to reduce this problem to black and white, the question of whether a general or flag officer is needed is ultimately subjective, and the final determination should be made by the services and joint organizations based on careful review of all objective and subjective factors.

Before defining and applying the factors, we begin with an overview of the current requirements for RC G/FOs: How many are needed? Of what rank? What kinds of positions do we need them to fill?

Overview of Current Requirements

The data we collected from the reserve components and joint organizations, which are summarized in Table 4.1, identified 634 requirements for an RC G/FO. Of note, the Chiefs of the Reserve for the Army and Air Force Reserve have to come from the general officers of the Army Reserve and the Air Force Reserve, respectively. So, even though the incumbents are on active duty, we consider these to be positions requiring a reserve component general officer. The Chief of the Navy Reserve, although not mandated by law, is traditionally a reserve component officer. The Commander Marine Forces Reserve is also not mandated by law, and is filled by either an active or reserve general officer (the current incumbent is active and the incoming nominee is reserve). For consistency, both are also included as positions requiring an RC G/FO.

Although the 634 positions detailed in Table 4.1 are the basis for our analysis, a complete picture of the positions that RC G/FOs may be called on to fill includes a number of other requirements that are not captured in this table—principally because they do not count against RC G/FO authorized strength. The additional uses for RC G/FOs include:

- RC G/FOs serving in active duty general and flag officer positions. As of November 2014, 12 RC G/FOs filled positions that count against active component

Table 4.1
RC G/FO Requirements, by Rank

Component	O-7	O-8	O-9	O-10	Total
Army National Guard	28	11	0	0	39
Army Reserve	83	44	1	0	128
Navy Reserve	32	20	1	0	53
Marine Corps Reserve	11	7	1	0	19
Air National Guard	29	9	1	0	39
Air Force Reserve	73	44	1	0	118
NGB	10	3	0	0	13
Joint organizations	38	22	2	1	63
TAGs and AAGs	108	54	0	0	162
Total	**412**	**214**	**7**	**1**	**634**

authorized strength. These requirements are not captured in Table 4.1 because these positions require a general or flag officer, but not specifically an RC G/FO. The assignment of RC G/FOs to these positions indicates that the active component has prioritized the use of its authorized strength for an RC G/FO over one of its own officers.

- RC G/FOs filling regular active duty joint pool positions. As of November 2014, 23 RC G/FOs filled joint pool positions. These requirements are not captured in Table 4.1 because the positions require a general or flag officer, but not specifically an RC G/FO. These are nominative positions, and the assignment of RC G/FOs indicates that the reserve component candidate was selected over all other nominees.

At the time we collected the data, the requirements plus the additional uses resulted in a total of approximately 670 RC G/FO positions. But these additional positions are not included in our analysis because the active component has established these requirements and authorized the positions, and active component authorized strength is used to fill the positions.

Further examination of the requirements for RC G/FOs reveals some interesting differences in how the services utilize these officers (Table 4.2). The Army Reserve, the Army National Guard, the Navy Reserve, and the Marine Corps Reserve use about 75 percent of their general and flag officers in command roles—specifically for commanders or deputy commanders. In contrast, only about 5 percent of the Air Force Reserve and Air National Guard general officer requirements are for commanders or deputy commanders. Instead, 80 percent of Air Force requirements are for assistants to senior leaders and commanders (either special assistants, National Guard assistants, or mobilization assistants). These variations reflect fundamental differences in how the services organize, employ, and mobilize their reserve components. In particular, it reflects the Air Force's use of associate units, which integrate active and reserve personnel in a single unit, in contrast to the stand-alone reserve units that predominate in the Army.

The state TAGs and AAGs are included in Table 4.2 for completeness, but they are not under the purview of the DoD and are not included in this review. The remainder of this chapter, therefore, focuses on the 472 remaining requirements. At the time the data were collected, 406 of the 472 positions were filled by general or flag officers; the remaining 66 were either vacant or filled by lower-ranking officers.

Factors for Evaluating General and Flag Officer Positions

To identify candidate positions for elimination, downgrading, or conversion, RAND researchers identified and applied a set of factors associated with general and flag officer requirements. Each position was tested against each factor in turn, filtering out positions that met one or more criteria. After testing all the key factors, what remained was

Table 4.2
RC G/FO Requirements, by Position

Component	Commander or Deputy	Director or Deputy	Chief of Staff	General Staff (J-code equiv.)	Special, National Guard, or Mobilization Assistant	Other	Total
Army National Guard	29	1	3	0	5	1	39
Army Reserve	102	8	8	5	4	1	128
Navy Reserve	37	10	1	3	2	0	53
Marine Corps Reserve	12	2	0	5	0	0	19
Air National Guard	2	3	1	0	33	0	39
Air Force Reserve	14	1	0	8	94	1	118
NGB	0	2	0	8	3	0	13
Joint organizations	13	3	1	28	18	0	63
Subtotal	209	30	14	57	159	3	472
State TAGs and AAGs							162
Total							**634**

a pool of positions that did not meet any of the filtering criteria and for which questions could reasonably be asked about whether the position needs to be filled by a general or flag officer. We then reviewed this pool.

A detailed review of each remaining position may identify further factors or combinations of factors that support a need for a general or flag officer. Such factors cannot be clearly identified for some positions, and those are the positions that we recommend the services review to ensure they are necessary (otherwise the requirement is a candidate for elimination) and if so, that the duties and responsibilities involved are commensurate with a general or flag officer (otherwise the requirement is a candidate for downgrading to O-6). Separately, we will explore the possibility of conversion to a senior civilian—SES, Defense Intelligence Senior Executive Service (DISES), or Defense Intelligence Senior Level (DISL).

RAND researchers selected seven factors for use in their evaluation:

- **Does the position have senior subordinates?** It is widely accepted that flag or general officers should serve as the superior of flag or general officers or SES civilians.[3] It is also commonly held that a flag or general officer should be the superior of O-6–level officers (especially when there are several), but this is more open to

[3] If the subordinates are not general or flag officers, but strictly SES civilians, then perhaps an SES member could take the place of a general or flag officer. There are six such situations; these are examined later in the analysis as part of the discussion of whether RC G/FO positions could be converted to senior civilian positions.

debate (and there are situations in which an O-6 is the superior of a small number of other O-6s).

- **Does the position exercise command of a military unit, or is it a deputy or vice commander of a military unit?** Command is a uniquely military role and the units that are appropriate for general or flag officer commanders are well established in service organizational structures. Any changes in which units are commanded by general or flag officers would require a servicewide review across both the active and reserve component organizational structures (e.g., which units should be commanded by general or flag officers), an issue that is beyond the scope of this study. Comparisons across services are not appropriate because of significant differences in how the services are manned and organized.
- **Does the position interact at a senior level with other organizations?** Other studies have consistently highlighted senior interactions as a primary factor associated with general and flag officer requirements. The general and flag officer rank provides stature, authority, and gravitas that are an important part of these interactions. In several cases, the reserve components have reported that O-6 officers cannot execute the responsibilities of the position because they are not allowed access into key meetings due to their lower rank. Given the part-time nature of many of the positions, and after discussions with OASD/RA, the services, and the Joint Staff, for the purposes of this filter, positions that dedicate at least 25 percent of their time to interacting at senior levels with any one of the following specified communities justifies a general or flag officer:
  - senior DoD (at the secretary, deputy secretary, under secretary, service secretary, and service under secretary levels)
  - senior Intelligence Community (director of national intelligence and agency director levels)
  - White House and National Security Council
  - Congress and congressional staff
  - other U.S. government departments (at the secretary and under secretary levels)
  - state governors and legislatures
  - foreign governments and militaries (at the ministry level, senior military leader level, or higher)
- **Does the position determine and set policy?** Shaping and setting policy, especially policy that is broadly applicable across larger parts of a service or across broad parts of DoD, is a role commonly accepted to require experienced leaders, such as general and flag officers, which is why prior studies included this factor. However, the determination of when policy issues are broad enough and important enough to merit a general or flag officer is subjective.
- **Does this position have the authority to negotiate commitments or international agreements with foreign nations on behalf of the United States?** Previ-

ous studies highlighted that the authority to obligate the United States internationally is closely associated with general or flag officers.

- **Is the position the director or deputy/vice director of an organization?** Prior studies associated directors of major organizations with a requirement for a general or flag officer. However, this filter is more contentious than the previous filters for two reasons: (1) the distinction between major organizations and smaller and/or lower-level organizations that may not require a leader at the general or flag officer level is subjective, and (2) an SES member may be able to serve as leader of an organization. We include this factor in our analysis, but later in this chapter we revisit the category of directors and deputy/vice directors as part of a review of positions that might be converted to senior civilian positions.
- **Is the position "new" since September 11, 2001?** "New" in this context means the position was either created, converted from active to reserve component, or fundamentally restructured subsequent to 9/11. This is the final and most debatable of the filters we apply. The arguments in favor of this filter are: (1) new requirements have gone through and passed internal review fairly recently; (2) the services have indicated the importance of the positions by making an explicit investment of limited resources in these positions; and (3) their current descriptions may not do them justice because the services need time to adapt and develop these positions into their ultimate roles. The arguments against using this as a filter are: (1) the post-9/11 needs that drove the creation of many of these new requirements are going away, and (2) many of these positions have been around for a decade—more than enough time for their roles to be fully developed and defined.

The analysis of which factors identify positions that should be filled with a flag or general officer has always been somewhat subjective and often debated. For this reason, our selection of factors built upon those that have been used and generally accepted in prior studies. Although we tried to identify a small number of generally accepted factors against which to filter general and flag officer positions, each factor still has some degree of debate (some more than others) and the precise interpretation of the factors may, in some cases, be subjective. To mitigate these concerns, we did the following:

- First, we applied the factor less stringently, so as to give the benefit of the doubt to accepting general and flag officer requirements. This approach produced a small set of positions that, even with a generous interpretation, do not meet any of the factors. While there is some risk that we may have "filtered out" positions with merit, this subset is worthy of review by the services.
- Then, we applied the factors much more stringently, eliminating some of the more-debated factors completely and being more selective in the application of the remaining factors. This produced a larger subset of positions, many of which

might be justified upon further inspection, but it allows us to explore and identify additional factors that are particularly relevant to RC G/FOs.

Table 4.3 shows which factors were used in each assessment. In the initial, less stringent filter, we used all seven factors. We dropped two factors in our more stringent filter. The first factor we dropped is "determine and set policy." For the less stringent filter, we interpreted this factor generously, giving credit to positions that perhaps have somewhat limited scope in their policy roles. By including this factor in the less stringent filter, we may retain some positions that have modest policy duties but no other factor associated with general and flag officers; the more stringent filter will catch these positions. The second factor that we eliminated in the stringent filter is the requirements that have emerged since September 11, 2001.

A third filter, "have general or flag officer, SES, or O-6 subordinates," is included in both filters, but the definition changes in the more stringent pass. In the less stringent filter, we assumed that positions with principle subordinates who are general and flag officers, SES, or O-6 need to be filled with a general or flag officer; the stringent filter limited this factor strictly to general and flag officer or SES subordinates.

Less Stringent Application of the Factors

Sequential application of the full group of factors, which we refer to as the less stringent set of filters, is presented in Figure 4.1, which shows the sequence in which factors were applied and how many positions were filtered out at each step of the process. The sequential application of each factor results in 27 positions, listed in Table 4.4, that do not meet any of the filtering criteria for staffing with a general or flag officer. However, there are other reasons why a position may require a general or flag officer that are not captured in the specific criteria we applied, so further review of these positions is appropriate. Potential reasons include the value of the position for development of general and flag officers, the need for senior-level perspective, and the scope and/or span of

Table 4.3
Factors Used in Evaluating Requirements for General or Flag Officers

Factor	Less Stringent Filter	Stringent Filter
Have general and flag officer, SES, or O-6 subordinates	X	X (general and flag officer or SES only)
Are commanders or deputy/vice commanders	X	X
Have senior-level interactions in 25 percent or more of their job	X	X
Determine and set policy	X	
Have the authority to obligate the U.S internationally	X	X
Are directors or deputy/vice directors	X	X
Are new requirements since September 11, 2001	X	

command and responsibilities. These reasons are highly subjective, depend greatly on context, and require individual review by the services.

One of the 27 positions serves as an example of why service review is so important. Using our factors, the Deputy Chief of Staff for Mobilization and Reserve Affairs, U.S. Army in Europe was identified as a candidate for elimination but stands out because it has responsibility over significant operational units and a large number of personnel and, therefore, deserves additional consideration. However, for the other 26 positions, the reserve components and joint organizations did not report any significant subordinates or any significant responsibilities over military units, military or civilian personnel, or significant fiscal resources. Also, none of these 27 positions was reported by the reserve components to have particular professional development value,

Figure 4.1
Less Stringent Application of Factors

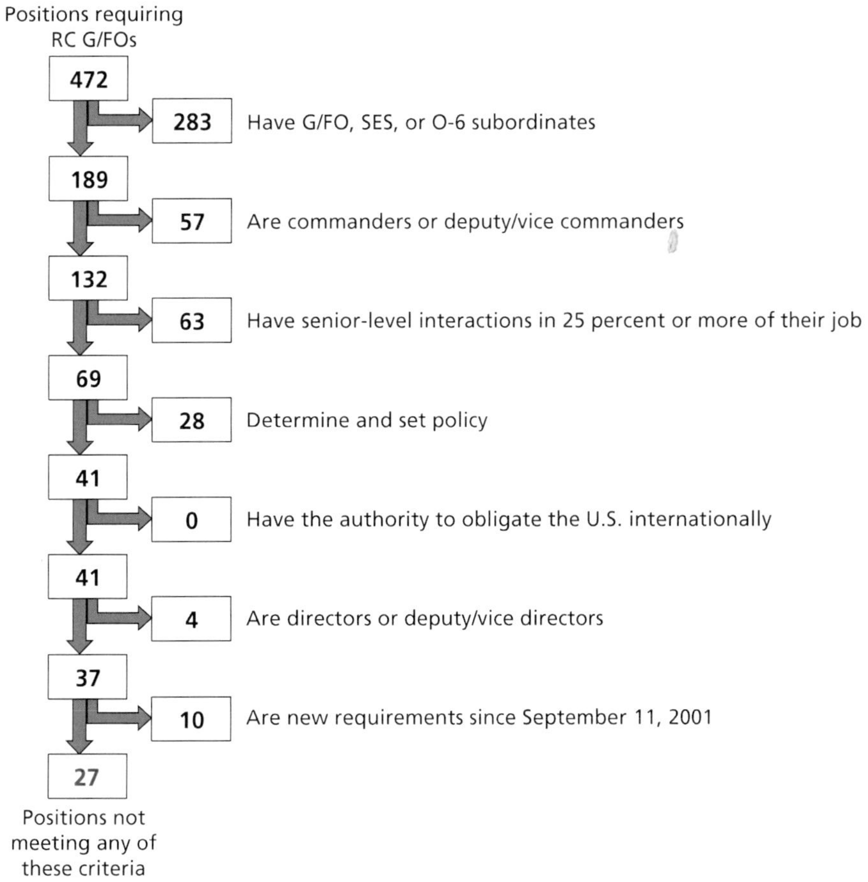

RAND *RR1156-4.1*

Table 4.4
Twenty-Seven Positions Remaining After Less Stringent Application of Factors

Duty Title	Organization	Location
U.S. Air Force Reserve		
Mobilization assistant to the commander	618th Air Operations Center (Tanker Airlift Control Center)	Scott Air Force Base, Illinois
Mobilization assistant to the commander	18th Air Force	Scott Air Force Base, Illinois
Mobilization assistant to the commander	U.S. Air Force Expeditionary Center	Joint Base McGuire-Dix-Lakehurst, New Jersey
Mobilization assistant to the commander	Air University	Maxwell Air Force Base, Alabama
Mobilization assistant to the commander	Curtis E. Lemay Center for Doctrine Development and Education	Maxwell Air Force Base, Alabama
Mobilization assistant to the commander	2nd Air Force	Keesler Air Force Base, Mississippi
Mobilization assistant to the commander	19th Air Force	Joint Base San Antonio, Texas
Mobilization assistant to the commander	8th Air Force	Barksdale Air Force Base, Louisiana
Mobilization assistant to the commander	20th Air Force	F. E. Warren Air Force Base, Wyoming
U.S Army Reserve		
Deputy chief of staff for mobilization and reserve affairs	U.S. Army Europe and 7th Army	Campbell Barracks, Germany
Air National Guard		
Special assistant to the director, Air National Guard for International Affairs	Director, Air National Guard	Arlington, Virginia
Assistant to the commander	U.S. Air Forces Central Command	Shaw Air Force Base, South Carolina
Assistant to the Surgeon General of the Air Force	Air Force Surgeon General	Pentagon, Arlington, Virginia
Assistant to the commander	Air Mobility Command	Scott Air Force Base, Illinois
Assistant to the commander	Air Education and Training Command	Joint Base San Antonio, Texas
Assistant to the commander	U.S. Air Forces in Europe	Ramstein Air Base, Germany
Assistant to the commander	Pacific Air Forces	Hickam Air Force Base, Hawaii
Assistant to the commander	Air Force Special Operations Command	Hurlburt Field, Florida
Assistant to the Command Surgeon	Air Combat Command, Command Surgeon	Langley Air Force Base, Virginia
Assistant to the Command Surgeon	Air Mobility Command, Command Surgeon	Scott Air Force Base, Illinois

Table 4.4—Continued

Duty Title	Organization	Location
Assistant to the commander	14th Air Force	Vandenberg Air Force Base, California
Assistant to the commander	18th Air Force	Scott Air Force Base, Illinois
Assistant to the director, Logistics, Installations and Mission Support, A4	Director, Air National Guard	Joint Base Andrews, Maryland
Assistant to the commander	U.S. Transportation Command	Scott Air Force Base, Illinois
Assistant to the Assistant Secretary of the Air Force, Manpower and Reserve Affairs	Office of the Assistant Secretary of the Air Force, Manpower and Reserve Affairs	Pentagon, Arlington, Virginia
Army National Guard		
Assistant to the Judge Advocate General	Army Judge Advocate General	Pentagon, Arlington, Virginia
Deputy chief of staff	Training and Doctrine Command	Fort Monroe, Virginia

except for several positions for which the Air Force Reserve reported that incumbents gain "numbered Air Force experience."

It is also noteworthy that only two of these 27 positions were vacant (as of February 2015). Given that 66 RC G/FO positions were vacant at that time, it suggests that the reserve components may value these positions more highly than our initial filtering implies. However, because we do not have data on the history of these vacancies over time—only a snapshot at the time of our data collection—we cannot draw definitive conclusions about the relationship between vacancies and requirements. Furthermore, vacancies do not necessarily occur because a position has a lower priority relative to others; they can occur for other personnel and operational reasons that have nothing to do with whether the position is truly needed by the service or joint organization.

The most notable feature of the 27 positions in question is that almost all of them are National Guard assistants or mobilization assistants. It is clear that some of these assistants function at very senior levels (e.g., the special assistant to the director, Air National Guard for International Affairs, and the Air National Guard assistants to U.S. Air Force Central Command, U.S. Air Forces in Europe, Pacific Air Forces, and Air Mobility Command). Other assistants are in positions that make them central players in planning and executing mobilization of the reserve components (e.g., the mobilization assistant to 618th Air Operations Center, Tanker Airlift Control Center and the Air National Guard assistant to the commander of U.S. Transportation Command). Other assistants serve at lower-level commands, where they also can play key roles in the mobilization of particular operational commands. However, it is unclear to us, especially without understanding plans for their roles once mobilization occurs, how to distinguish when mobilization assistants and National Guard assistants at the general officer level are appropriate, and when lower ranking officers can satisfactorily perform these roles. We will have more to say about mobilization and National Guard assistants later in this chapter.

Stringent Application of the Factors

Our less stringent application of the factors produced 27 positions that merit further review and consideration by the services. However, it is possible that we were too generous and that more positions deserve closer inspection. The most debatable elements of the initial filtering scheme are the following:

- O-6s do not always need a flag or general officer to serve as their superior; the filter should exclude only positions with flag, general, or SES principle subordinates.
- What constitutes *broad policy* is highly subjective, so this factor should not be used for filtering; it should be considered subjectively when examining the results of filtering.
- New positions need to be examined on the same basis as more-established positions; they should not be filtered out *a priori*.

We addressed these concerns in our more stringent application of the factors. First, O-6 subordinates are no longer sufficient to trigger the factor associated with senior subordinates. Second, we eliminated two filters: positions having a policy role and new positions put in place since 9/11, as Table 4.3 indicated.

Application of the remaining factors in sequence produced the cascade shown in Figure 4.2, which resulted in 95 positions that do not meet any of the criteria (see Appendix G for a complete list).

Because the filter criteria have been significantly narrowed, we expected this larger list to contain positions that, upon closer inspection, may have reasonable rationale for being filled with flag or general officers—and we found this to be the case. Simply reviewing the titles of those 95 positions reveals the following:

- Seven single-digit J-code generals of the NGB appear on the list; these are clearly appropriate positions for general officers.
- Various chiefs of staff and senior staff officers also appear on the list (e.g., several assistant commandants, Defense Logistics Agency J-9, and Chief Information Officer/G-6 Cyber Security Director). Positions such as these are commonly filled with general or flag officers, although each case requires individual review.
- Several positions in specialized communities appear on the list (e.g., the Chief of Judges of U.S. Army Legal Services, the Assistant and Deputy Chiefs of Chaplains, and Assistant and Deputy Surgeons General). The scope and level of responsibilities of these positions are broadly commensurate with those of general and flag officers in analogous roles.

Further review of the 95 positions also revealed the following:

- More than two dozen of the positions are reported by the reserve components as developing skills and/or providing experience needed for the development of more-senior general and flag officers. While some of the positions identify skills in generic terms (e.g., "joint experience" or "numbered Air Force experience"), many of the two dozen positions identify relatively specific skills and experience (e.g., "ability to lead in multi-layer, multi-functional environment"; "comprehensive joint, inter-agency, inter-government, multi-national experience"; "mastery of current USG [U.S. government] and DoD policy for nine separate and diverse Unified Command Plan . . . missions"). It is difficult to determine how much weight professional development should carry in setting general and flag officer requirements (see Chapter Five), and it is even harder to ascribe weight to the more-generic statements, but the data do indicate that some of these positions have specific developmental value.
- Responsibility over personnel is a very subjective criterion when considering whether a general or flag officer is required. In operational units, it is common for O-5s to have responsibility over hundreds of personnel and O-6s over thousands (and this varies greatly between services). However, in headquarters organizations, many general and flag officers have responsibility over no more than

Figure 4.2
Stringent Application of Factors

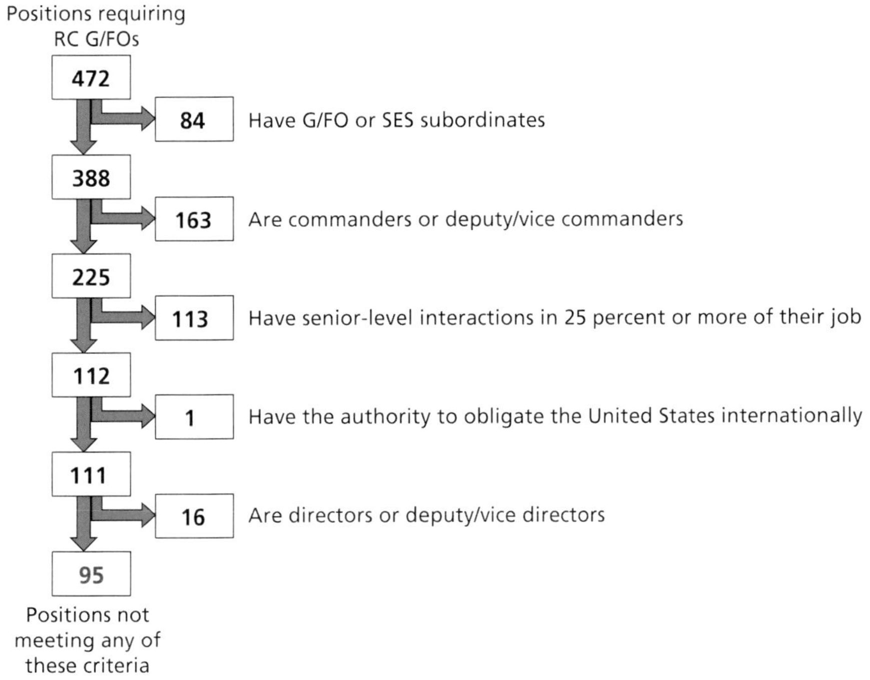

RAND *RR1156-4.2*

several dozen people. Nevertheless, this matters at some level. The reserve components and joint organizations reported that about a dozen of the 95 positions have responsibility for between a few thousand to 200,000 personnel. Depending on the reserve component and nature of the position (e.g., headquarters staff or unit command), responsibility over personnel would be a reasonable consideration in requiring that a general or flag officer fill the position.

- Similarly, the magnitude of the fiscal responsibilities is also a very subjective criterion when considering whether a general or flag officer is required, and can depend greatly on the service, the nature of the unit, and particular role of the position. Nevertheless, this also matters at some level. The reserve components reported that seven of the 95 positions have significant obligation authorities, ranging from a few hundreds of millions of dollars to billions of dollars. Depending on the service and the nature of the organization, fiscal responsibility could reasonably play in the decision to require a general or flag officer.
- In some cases, reserve and joint organizations had blank entries in their data call responses where, based on the type of position, one might have expected data that would have filtered out the position. More-detailed review may show that this is a case of incomplete reporting.

Collectively, these considerations suggest that perhaps half of the 95 positions have reasonable arguments for requiring general and flag officers. Precise determination will depend on highly contextual and subjective aspects of each position that we deferred to service review.

What is most notable about the subset of 95 positions is that, like the previously identified set of 27, most of these positions are special, National Guard, and mobilization assistants, as shown in Table 4.5. This distinct pattern led us to take a specific look at such assistants, which we do in the next section.

Special, National Guard, and Mobilization Assistant Positions

As shown in Table 4.2, 159 of the 472 RC G/FO requirements are for special assistants, National Guard assistants, or mobilization assistants. In reviewing the position descriptions, individuals in positions serve dual purposes: to inform and advise the commands on the capabilities and employment of reserve and National Guard forces, and to keep the parent reserve component informed of the activities and needs of the commands. The depth and scope of this role is generally not clear from the information we obtained, and we expect it to vary considerably according to the nature of the command.

Furthermore, the responsibilities of these assistants in times of mobilization are unclear. At higher-level headquarters, it is possible that these assistants would remain in an advisory, liaison, and/or coordination role; at more-operational headquarters, we would expect that individuals in some of these positions would take on leader-

Table 4.5
Positions Remaining After Stringent Application of Factors

Component	Chief of Staff	General Staff (J-code equiv.)	Special, National Guard, or Mobilization Assistant	Other	Total
Army National Guard	2	0	5	1	8
Army Reserve	5	5	2	1	13
Navy Reserve	0	3	1	0	4
Marine Corps Reserve	0	5	0	0	5
Air National Guard	0	0	33	0	33
Air Force Reserve	0	0	9	0	9
NGB	0	7	3	0	10
Joint organizations	0	3	10	0	13
Total	**7**	**23**	**63**	**2**	**95**

ship responsibilities in actual operations. However, the actual mobilization role is not clearly specified in the individual position information provided to us, and it is likewise not described in any formal policy document at the DoD, joint, or service level. This makes it very difficult to independently assess whether all of these assistant positions are appropriate for general or flag officers.

Further complicating the issue, these positions are closely intertwined with the role of individual mobilization augmentees (IMA). IMAs fill most of the special and mobilization assistant positions (106 of 159). In fact, more than half of all IMAs are used to fill assistant positions. Neither the exact nature of an IMA's liaison role nor an IMA's role in mobilization and wartime is clearly specified in any document that we know of.[4] This makes it difficult for us to systematically understand the reasons that these positions require general officers.

The loosely defined status of the IMAs and the difficulty it causes in clearly establishing requirements for general officers was highlighted as long ago as the Hay Group's 1992 *Evaluation of Reserve General and Flag Officer Positions*. This study stated:

[4] DoD Instruction (DoDI) 1235.11 (DoD, 2015c) provides very general guidance on the management of IMAs. According to the instruction, IMAs must be used to "support mobilization requirements, contingency operations, operations other than war, or other specialized or technical requirements." DoDI 1235.11 does not specifically address the appropriate roles for IMAs at the rank of general or flag officer. Air Force Instruction (AFI) 36-2629 (Department of the Air Force, 2012) also very generally addresses the purpose of IMAs as force multipliers in "war, contingency operations and peacetime to meet National Defense, strategic national interest, and domestic objectives" primarily to support their unit of assignment. AFI 36-2629 does not specifically address IMAs at the rank of general officer, except to assign responsibility for validating and tracking mobilization assistant positions to the Chief of the Air Force Reserve. Army Regulation 140-145 (Department of the Army, 2012) establishes procedures for employing, using, and managing IMAs, but does not address appropriate roles for general officer IMAs.

> The functions and pattern of utilization of Individual Mobilization Augmentees (IMA) created a number of problems for the job evaluation panel. [. . .] In some cases they are clearly involved as a senior advisor and have a defined wartime mission. In these cases the panel was able to completely evaluate the job. In other cases there is not a well-defined wartime mission. In discussing these cases the panel determined that the unpredictability of the potential contingencies and the varied skills of senior G/FO [general and flag officers] officers led to a situation where specific criteria upon which to base hard and fast evaluations regarding IMAs could not be developed. However, the panel felt that in a full mobilization demand for skills would exceed the supply of talented officers and that roles would be found for the officers or they would not be activated. In those cases where the wartime role was not specifically defined, the panel evaluated the jobs based on the dual role of advisor to the active force and the reserve component, which required significant know-how even though the other components of the job could not in all cases be accurately gauged. *In Hay's view, these assignees without a clear wartime role need careful monitoring to ensure they are being fully utilized in order to justify these positions at the G/FO level.* [emphasis added]

We are in general agreement with the Hay recommendation and further suggest that the services set standards and expectations for the National Guard assistant and mobilization assistant positions that define the nature of their roles and responsibilities.

Positions Created Since 9/11

Of the 472 positions requiring a general or flag officer, 96 are new requirements established since September 11, 2001. These include 90 positions that are completely new, four positions that were realigned from the active to the reserve component, and two positions that resulted from a command restructuring that occurred post-9/11. These 96 positions cover a wide spectrum, with more than half of the new positions being commanders, deputy commanders, and positions on general staffs, as can be seen in Figure 4.3.

The new positions include some very senior positions (e.g., the vice chief of the National Guard). About 15 percent of the new positions are in operational units (almost all of the operational positions being commander or deputy) or positions in theater (such as at International Security Assistance Force, U.S. Forces Afghanistan, or Combined Joint Task Force–Horn of Africa). Eighty percent of the positions are at major standing headquarters (42 percent at joint headquarters, 38 percent at service headquarters). The remaining new positions are at organizations dedicated to support or training functions. This distribution is not surprising, given the post-9/11 demands on the total force. Even new positions for assistants (be they special, National Guard, or mobilization) correlate very directly to the changes that have occurred in the force since 9/11, with half of the new assistant positions located at combatant commands, service commands, and joint organizations. The other half are located at service chief

Figure 4.3
New Requirements for RC G/FOs Since 9/11, by Position

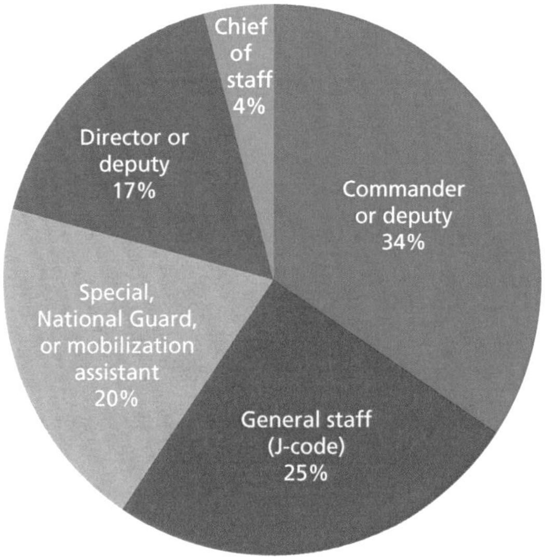

RAND *RR1156-4.3*

headquarters working at senior levels (e.g., the A2, A5, A6, and A7 assistants to the director, Air National Guard, and the special assistant to the Assistant Secretary of the Army for Manpower and Reserve Affairs).

Overall, this indicates that the new requirements align with the overall emerging requirements since 9/11. These past 15 years have seen long-term sustained operations in two joint operations areas, two separate troop surges, and high demands for specialties that, in many cases, were mostly available in the reserve components (e.g., medical units; mission support elements; construction units; intelligence, surveillance, and reconnaissance units; and signals units). It is also noteworthy that eight of the new requirements are directly related to homeland defense missions that directly involve both active component and National Guard elements.

However, as activities in the two joint operations areas evolve, it makes sense for the services to review these requirements to determine which are enduring. Many of them, such as the new requirements related to the establishment of the Chief of National Guard as a member of the Joint Chiefs and those related to the increased emphasis placed on homeland defense missions, will clearly be enduring. But those new positions directly tied to current forward operations (beginning with the new requirements that are actually in theater) should be reviewed as part of the drawdown and evolution of these operations. The reduction in funding for overseas contingency operations also provides a natural opportunity to review the status of general and flag officer requirements being paid for using those funds.

The fact that 80 percent of the new requirements are at headquarters staffs is of interest, especially in light of the recent efforts to reduce the size of headquarters staffs. The growth in requirements for reserve component positions at these commands may be a response to the downsizing of the active military and civilian components at many headquarters. However, given that continued pressure to reduce staff sizes is expected from Congress, the Secretary, and the Deputy Secretary, the services would do well to scrutinize and review the growth of RC G/FO positions at headquarters staffs to ensure that they are enduring requirements and that they do not put the reserve components at risk of inadvertently undermining senior leader efforts to downsize headquarters staffs.

Conversion to Senior Civilian Positions

Another question raised by Congress is whether a senior civilian could fill any of the current RC G/FO requirements. We obtained from the services several position descriptions for SES positions, but when we compared these with general and flag officer position descriptions, we had difficulty finding factors that would help us distinguish between the two. So, we chose instead to identify a sample of RC G/FO positions likely to contain potential candidates for conversion to SES and examine these positions more thoroughly. The sample we identified for review consists of positions that either

- are directors or deputy directors
- have SES members but not general or flag officers as principle subordinates.

There are 34 such positions, of which 27 are filled by RC G/FOs that are either IMAs or traditional reservists who perform these duties on a part-time basis. Part-time assignments are incompatible with SES positions, which are almost exclusively full-time. Of the remaining seven positions, three clearly should be military:

- director, Reserve Affairs, U.S. Marine Corps
- director, Air National Guard
- deputy director, Air National Guard.

Two other positions were explicitly designated as Chairman's Reserve Positions and filled with reserve component officers activated for two-year terms. The remaining three are perhaps less clear:

- director, legislative liaison, NGB
- director, Army National Guard for Aviation Transformation
- deputy director, Intelligence and Knowledge Development (J2) at U.S. Africa Command.

Traditionally, the services use uniformed personnel as the chief of their legislative liaison offices. The second position, aviation transformation, may be most effectively led by a person experienced in how the Air National Guard utilizes aviation, although perhaps an SES member that is retired from active duty with appropriate experience might be able to perform the job. The third position has the greatest potential for SES conversion, as it has no general or flag officers as subordinates and other combatant commands have deputy directors of intelligence that are SES equivalents from the intelligence community (DISESs or DISLs), although a definitive determination requires a detailed position analysis that is beyond the scope of this review. This position is highlighted for having value in developing the expertise of RC G/FOs in working with allied and interagency communities on intelligence and counterintelligence issues, so any potential conversion to DISES/DISL would have to be weighed against the loss of a development opportunity for the RC G/FO community.

This last position also raises the larger question of whether general or flag officers—in both the active and reserve component—on headquarters staffs can be replaced with SES (or equivalent DISES/DISL) civilians. A number of high-level staffs, such as the Joint Staff and combatant command staffs, have placed SES personnel into senior leadership positions. For instance, the J-8s at U.S. Central Command, U.S. Africa Command, and U.S. Pacific Command and the Deputy J-2 at U.S. Central Command are all members of the SES. In our experience, such conversions are highly dependent on the particular situation and context, making it difficult—if not impossible—to establish overarching rules and criteria. Broadly speaking, whatever makes a general staff position convertible to SES should not depend on whether the position is in the active or reserve component, with two exceptions: (1) part-time RC G/FO positions cannot be converted to SES positions, and (2) the developmental value to the reserve component community should be considered.

In conclusion, our review of positions that could most likely be converted to SES found only one or two positions that *might* merit further review. For this reason, we believe such opportunities are very limited. The larger question of whether any general or flag officer position on headquarters general staffs can or should be converted to SES positions is a complex and contentious issue encompassing general and flag officers in the active and reserve component and is beyond the scope of this review.

Findings and Recommendations

In response to the congressional language inquiring about the potential for downgrading, eliminating, or converting of the current requirements for RC G/FOs, we collected data on these requirements from the reserve components and joint organizations. Upon reviewing these data, we reached the following conclusions.

The number of RC G/FO requirements exceeds authorized strength. The total number of requirements in combination with other roles for which RC G/FOs are needed is somewhat larger than the available authorized strength. As of February 2015, 66 positions requiring RC G/FOs were vacant, due to either the limitations on RC G/FO authorized end strength or the lack of funding.

We see some potential opportunities for eliminating or downgrading RC G/FO requirements. Our initial filtering identified only 27 of 634 positions (about 4 percent) as candidates for further review by the services to revalidate them as general and flag officer requirements, or consider them for elimination or downgrade. Important contextual, qualitative, and subjective factors make it impossible for us to make purely analytical determinations on these 27 positions. Some of the 27 positions stand out fairly clearly as reasonable positions for general and flag officers, and service review may indeed revalidate all or most of these. A second, more stringent filtering produced 95 positions, but review of some additional factors suggests that about half of these (which include the original 27) might be worthy of detailed review for possible reductions.

There are few opportunities to convert RC G/FO requirements to senior civilian positions. The biggest impediment is that many of the RC G/FO positions are part-time, although the value of the positions for officer professional development may also be an important consideration. There is a separate issue (affecting both active and reserve components) of whether some general and flag officer positions on headquarters general staffs (e.g., at a combatant command headquarters, service chief headquarters, or major command headquarters) can or should be converted to senior civilian positions—though answering this question is beyond the scope of this review. While this larger question affects both active and reserve positions, we wish to highlight that 65 percent of the reserve component positions on general staffs are identified as having important value for the professional development of RC G/FOs.

Positions described as mobilization or special assistants (most of which are filled by IMAs) or National Guard assistants are the requirements category that raises the most questions. These positions have loosely defined peacetime roles that can be generically described as liaisons between their commands and the reserve components. Their roles upon mobilization are not well articulated. As these constitute one-third of the 472 general and flag officer requirements, it would help justify these positions and assuage congressional concerns if policies and standards existed that defined the expected roles of these positions both in peacetime and during mobilization.

RC G/FO requirements that have emerged since 9/11 broadly reflect the needs of the past 15 years, but should be reviewed as operational requirements change. Many of the RC G/FO requirements added since 9/11 are closely related to enduring changes in the force structure and organizations, but others are more closely tied to operational requirements that continue to evolve. These requirements should be reviewed on a regular basis to ensure that limited general and flag officer resources are dedicated to the highest priority missions.

These conclusions lead us to make the following recommendations:

- **The services and joint organizations should review the 27 positions identified in our analysis to ensure these remain valid RC G/FO requirements.** Those that are no longer valid should be considered for elimination or downgrading.
- **OUSD(P&R) should assist the services in establishing overarching guidance on how RC G/FO IMAs should be used and the roles they should play.** While flexibility to tailor assignments is essential, there should be a common foundation that establishes the basis for how all the services employ RC G/FO IMAs.
- **OUSD(P&R) should assist the services in establishing guidance regarding the overall nature of positions described as mobilization assistants or National Guard assistants.** There should be a common understanding of the duties and responsibilities associated with such positions both in peacetime and at times of mobilization. Such guidance should assist the services in validating and justifying such positions.
- **The services and joint organizations should review RC G/FO requirements established during the past 15 years.** These include some that are clearly enduring, but also a subset that is more closely tied to operational requirements in the Central Command area of responsibility. As those operations end or evolve, there may be opportunities to free up valuable RC G/FO resources for use elsewhere.

CHAPTER FIVE

Considering Development in Setting Requirements

Past reviews of general and flag officer positions have not included officer development as a criterion for assessing requirements. There are several likely reasons for this. Most general and flag officers have at least a quarter-century of service behind them and are viewed as being very well developed at that point in their careers. Professional development is a personal and idiosyncratic process; it is not easy to consider the developmental benefit of a position without considering how it fits into a particular officer's career path. What may be a new challenge to one officer is old hat to another. The payoff for assigning an officer to a developmental position—or the loss for not making such an assignment—can also be speculative and occurs at some point in the future. Military command, resource management, and senior-level interactions are more tangible and urgent, and thus make it far easier to justify the need for a general or flag officer position.

This is not to say that development is not a worthwhile consideration for general and flag officers. Such officers may serve a decade or more after promotion from O-6, and it is reasonable to expect them to continue to grow and develop, given the magnitude of their responsibilities. Because professional development depends so much on work experience, which is gained primarily through job assignments, we consider a position's developmental value as a criterion for this study, alongside the nature and magnitude of responsibilities discussed in other chapters. In theory, by demonstrating that some jobs are essential to cultivating a high-quality senior officer cadre, development issues could provide an important lens on the question of whether the services have the appropriate number and type of general and flag officer positions.

We set out to address these questions by evaluating whether development considerations might be useful in determining overall RC G/FO requirements. We conducted a literature review of available research on general and flag officer careers, leader development, strategic human resource management, learning by experience, and related areas. We performed empirical data analysis of representative sets of RC G/FO positions and identified patterns. And we conducted informal conversations with current and retired RC G/FOs to acquire information on the relationship between development and requirements. This part of our research did not represent an examination of officer development *per se*, but a consideration of the uses of development as a lens to assess general and flag officer requirements.

Defining Development

The concept of development has a wide and complex array of potential meanings. One fundamental distinction is between individual-level development (programs designed to maximize the talents of specific leaders) and institutionally focused development (a process to generate the best qualified slate of candidates for progressively more-senior positions). In individual terms, for example, development scholar Cynthia McCauley has argued that leader development involves programs to develop "a person's capacity to be effective in leadership roles and processes" (McCauley and Brutus, 1998), and the parallel effort to provide leaders with opportunities to grow and learn. Leadership development in a more institutional sense is about building "benches"—a sufficient number of highly qualified leaders to provide appropriate selectivity for higher positions.

These two basic approaches need not be mutually exclusive, of course. One source defines management development as "an organization's conscious effort to provide its managers (and potential managers) with opportunities to learn, grow, and change, in hopes of producing over the long term a cadre of managers with the skills necessary to function effectively in the organization" (McCall, Lombardo, and Morrison, 1988, pp. 147–148). That is to say, organizations frequently use individual development as a means to the end of furthering institutional goals (McCall, 2004, p. 129).[1] And this is precisely the case with U.S. military services that view their developmental investment in individuals as part of an effort to build a strong institutional bench.

Measuring Developmental Value

Empirical evidence suggests that experience constitutes the most powerful and effective developmental tool—more so than education, training, or mentoring. McCall concludes that, "The primary source of learning to lead, to the extent that leadership can be learned, is experience"(McCall, 2004; Campion, Cheraskin, and Stevens, 1994; McGuire, 2001–2002; McCall, Lombardo, and Morrison, 1988, p. 1). Training, education, and mentoring can be tremendously important, but should be supportive of experience (McCauley and Brutus, 1998, p. 4). A conscious and coherent approach to using successive jobs as levers for officer development therefore enjoys strong support from the existing research.

---

[1] In a similar fashion, Morgan McCall (2004, p. 129), one of the chief scholars in the field, has argued that the primary objective of a development program is to ensure "that people in leadership roles have the competence to determine and to carry out the strategic imperatives" of the organization. This again marries the two basic approaches, suggesting that individual development is a route to institutional effectiveness.

Certain categories of positions do appear to carry disproportionate developmental value. The literature suggests, as one source concludes, that: "Not all experiences are created equal. . . . Some experiences simply pack more developmental wallop than others" (McCall, Lombardo, and Morrison, 1988, p. 5). Most learning and new conceptual capacity, for example, emerges when "leaders are pushed beyond their current frame of reference" (McGuire, 2001–2002, p. 93) and jobs that tend to do that—ones that offer fresh, different, demanding experiences—are believed to have greater developmental effects than continued experience along a well-worn path. Repeated similar experiences, on the other hand, carry diminishing value for development. Experiences that offer the most potential for development tend to fall into a number of categories—jobs that deal with:

1. unfamiliar, new issues, including working with other organizations and different cultures, great variety
2. leading change, including dealing with bad conditions, starting an initiative from scratch, executing a turnaround
3. high-responsibility tasks with wide latitude—jobs with big "stakes," both organizationally and personally, and wide scope, in terms of number of staff and geographical and functional dispersion
4. requirements to work across organizational boundaries or with outside organizations, using influence rather than authority
5. diverse sets of people, including running a team or task force of multiple functions, skills, and specialties, as well as dealing with other people through negotiations and persuasion
6. intellectually challenging issues, with a substantial strategic component
7. senior leaders that are multiple levels above someone's current position.

These categories point to an important finding: The general concept of "broadening" is integrally related to senior leader development. The essence of senior leadership is cultivating an enterprise-level vision and transcending technical or operational mindsets to a comprehensive view of responsibilities and challenges. Studies suggest that the most common reason for executive failure is the inability to make this shift in perspective and learn to operate in a more encompassing domain. Positions that inherently reflect such a broadening experience—as well as those that represent other criteria on the list above—therefore potentially offer the greatest developmental value to organizations.

Limits to Developmental Analysis

The conclusion that some jobs are inherently more developmental than others comes with three important caveats. First, the relationship between experience and development tends to be contingent rather than universal. Put simply, a given experience will be more developmental for some leaders than others. Peoples' interaction with experiences is subjective; the key is not the job itself but the job *as experienced by an individual leader*. The same situation can have very different developmental effects, and different people learn differently from the same situation (McCauley and Brutus, 1998, pp. 9, 44). As McCall (2004, p. 128) concludes,

> The challenge in using experience for development lies in giving the right experiences to the people who will learn the most from them . . . and then providing the kind of support that will help them learn what the experiences offer. This is anything but easy. It requires developing ways to identify what experiences are developmental and where they are, ways to identify the people with the ability to learn from those experiences, mechanisms for getting the right people into those experiences, . . . ways to identify and specify desired learning outcomes, and an understanding of the kinds of interventions that promote the developmental side of performance-driven assignments.

Some people are simply better at learning from experience than others. Research suggests that key determinants of learning from experience include learning orientation, openness, critical reflection, and a proactive stance toward problems and opportunities (McCauley and Brutus, 1998, p. 48). The key to success in development is not merely finding the right experience, but putting people in developmental positions and then helping them learn through structured rather than arbitrary experience. That is one reason why developing a coherent philosophy for development turns out to be so difficult—it helps to guide the choice of priorities, the ways an organization will balance the multiple competing goals at work in a development process (Thie et al., 2001).[2] As a result, even jobs with high developmental potential in theory may not have the desired effect in practice.

Second, experience itself does not always maximize the learning potential of a given job assignment. The literature suggests the value not merely of experience but of "structured" experience—efforts to shape the way leaders encounter the developmental opportunities to make them most effective (Jacobsen, 2007). These issues require more attention because "the process of learning from experience . . . is largely unmanaged and unmeasured" (Dean and Shanley, 2006, p. 4). In civilian corporations as well as the military services, despite the importance of structured experience to development, very little effort is put into maximizing the return on investment for experiential learning; it has been "seat of the pants" rather than strategic (McCall, Lombardo, and Morrison, 1988, pp. 2, 5). Numerous studies, for example, point to reflection as a critical adjunct to learning from experience, but it is seldom required or even enabled (Dean and Shanley, 2006, pp. 3, 5–6), (McGuire, 2002, p. 96).

---

[2] As one example, a study that examined the effects of longer tenure suggested that it would enhance stability, if that was the goal—but would have little effect on other goals.

Our discussions with the military departments' general and flag officer management offices make clear that this gap exists even amid very well developed and carefully thought-through general and flag officer development programs. All the services work to develop concepts for leader development specific to each general or flag officer, and all offer some form of education or training to prepare senior leaders for their new responsibilities. Once the general and flag officers are in the job, however, they are largely on their own to ensure that they get the best possible learning value from the experience.

This is not to suggest that the services do not have development frameworks in place. In fact many do, and several are in the process of building creative new procedures. But there are fewer designated development activities once officers cross the O-7 threshold: Development becomes a far more individualized, and to some degree haphazard, process as the leaders of reserve components make a persistent series of judgments about filling positions that come open, and using various jobs as developmental experiences. The use of individual development plans becomes much less common at the general and flag officer level.

This points to a third complication in developing general and flag officers—the inherent nature of their assignments. The general and flag officer assignment process is shielded, personalized, and idiosyncratic. It reflects a dynamic, continual process of matching well-suited leaders to specific jobs under the pressure of a dozen or more constraints: the need to fill open positions, matches between skills and job requirements, officers' broadening needs, timing, personalities, the sense of a specific officer's future trajectory, and many more. No static framework for development will be able to capture such a complex and often unpredictable array of variables. Matching can take place only through in-depth dialogues about the needs and opportunities for the institution, the specific competencies and capabilities of the officers who might fill the positions, and the future benefits that such an assignment might accrue to the candidate officers and the services—the developmental value of the assignment.

Using Development to Evaluate the Adequacy of RC G/FO Requirements

To better understand the developmental value of positions, we talked to general and flag officer management offices about development philosophies, and we looked at how general and flag officers actually move through different positions during the course of their careers.

Service Development Philosophies

Several themes emerged from our discussions with the general and flag officer management offices. First, general and flag officer development programs are overwhelmingly oriented to institutional leadership development goals. Individual development plans are means to institutional ends, not the end itself. General and flag officer management offices focus on building the most qualified and capable slate of officers for progressively senior positions—and, even more specifically, for the senior-most ranks in their chain of advancement. The *CJCS Vision for Joint Officer Development* from 2005 says it simply: The objective of officer development is "to produce the largest body of fully qualified and inherently joint officers suitable for joint command and staff responsibilities" (CJCS, 2005, p. 3). This emphasis does not ignore the role of individual leader development: It is by offering particular officers interesting, broadening experiences that a service will create a sufficient and competitive pool of future candidates. But the focus is on producing the strongest bench for higher positions.

Second, the services do appear to regard certain general and flag officer positions as "developmental," but only in a very broad sense. Broadening—giving officers a wide range of experiences across various institutional functions—is the basic tool or mechanism to build a sufficient bench of officers qualified for more-senior jobs, leaders who have been equipped with the range of experiences viewed as essential to those positions. Some jobs (such as positions in joint environments and command headquarters) are typically viewed as more developmental than others, but there is little appetite to protect, expand, or eliminate certain types of positions *because of* their developmental role.

Therefore, the services tend to categorize as especially developmental those jobs that provide essential rounding experiences in several categories. Command at the general and flag officer level is one category, and essential to more-senior leadership. A second is staff jobs that deal with programming, finance, and budgeting issues, which account for a growing proportion of a general or flag officer's time and responsibilities. Third, all the services place emphasis on the developmental role of assignments outside the normal service channels of operational advancement—positions beyond operational units, such as jobs at combatant commands, service headquarters, and especially in joint environments. All the services look at officer assignments with an eye toward creating well-rounded officers who can arrive at senior O-8 jobs (or, in the reserve components, compete for the single O-9 post) having covered a number of these critical bases.

A partial exception to this typical practice—but only partial—is the way the services conceive of development for particular specialties or areas of domain expertise, such as signals, medicine, law, intelligence, and logistics. In these and similar areas, domain-specific knowledge is highly valued and its cultivation becomes an important aspect of developmental plans. Being qualified for more-senior positions in such areas is often partly about possessing specific domain expertise to operate effectively in that environment. Even in such cases, however, all the services expressed a commitment to the goal of creating well-rounded general and flag officers with a wide range of experiences. The number of jobs for which the development process involves mostly the accumulation of highly discrete domain expertise is very small; in fact, our research did not uncover any RC G/FO jobs that fit such a description.

A fourth category is that this pattern—the absence of simple, singular developmental paths—emerged in our empirical review of officer biographies to determine whether career paths offered any evidence for specific developmental "channels" or lines of advancement. Officer development does not appear to follow clear, identifiable channels through a well-established pattern of jobs. Officers who had held one O-7 job did not strongly tend to hold a few specific O-8 jobs afterward; patterns of advancement varied widely and allowed many paths through a very large number of possible combinations. (The exceptions, again, are in narrow fields where domain expertise is required for a few specific positions.) The services appear to be trying to assemble a broadening set of experiences on a leader-by-leader basis.

Empirical Analysis of Developmental Benefits of RC G/FO Positions

In Chapter Four, we presented a methodology to evaluate the validity of RC G/FO requirements. This analysis suggests that roughly 85 to 95 percent of current RC G/FO positions meet our criteria for justification. Those positions may confer some developmental benefit upon their incumbents, but there is no need to justify a requirement on that basis when it is already justified for some other reason. For the remaining 5 to 15 percent of the positions, however, the developmental considerations are singularly important: A position may be worth keeping, or even necessary, if it is regularly used to prepare general and flag officers for positions with greater responsibility.

As the preceding discussion makes clear, discerning developmental benefits is more art than science—since one person's routine assignment could be another person's developmental assignment. Almost any assignment can be developmental in the sense that an officer learns something from doing it and carries that learning to the next assignment. But by that standard, no positions at all would be eliminated. Rather than offer philosophical arguments about the developmental benefits of particular positions, our analysis seeks evidence that positions are used to develop officers *in practice*. As a practical matter, it is easier to identify positions that are *not* used developmentally than to identify ones that are. For example, if a position tends to be the last one a general or flag officer holds before retiring, or if it—at best—precedes lateral assignments without subsequent promotion, we might conclude that it is not being used for a developmental purpose.

The test we are applying, then, does not attempt to measure the potential developmental value of a position; rather, we seek evidence that it is being effectively used for developmental purposes by the service. In theory, some positions that we identified as being poorly employed to serve developmental goals could be better employed. However, if the services are not regularly employing a position to develop general and flag officers for higher positions, it would be very difficult to make the case for its necessity on developmental grounds, given that all services reported sufficient positions to achieve necessary development goals.

The criterion we employed does leave room for interpretation. Chance could be a factor. Services might assign general and flag officers to these positions as developmental experiences with every intention of promoting them, and our findings might be simply uncovering the random fact that, for two or three general or flag officers in a row, expected or hoped-for promotions did not happen. For this and other reasons, these findings should be taken as suggestive rather than conclusive. They point to the need for further analysis of a number of positions that do not seem to be playing a significant developmental role.

Of the 95 positions that were not captured by the other filters to justify a general or flag officer position, we rule out as developmental the 34 positions that are coded for O-8. This is the highest grade that all but a handful of reserve component officers achieve, and it is unlikely that officers in their terminal grade will be given an assignment for developmental reasons, except in unique circumstances. This does not imply that these positions are not needed, only that there are few if any cases in which they can be justified on developmental grounds. That still leaves 61 O-7 positions for further analysis.

The following analysis examines the role of these 61 positions (in the context of the total 95) in the career progression of general and flag officers. To conduct this analysis, we examined the biographies of active and retired general and flag officers from each of the reserve components dating back to September 2001. An important caveat is that our analysis is only as good as the quality of the information on the officers' public biographies. We found instances where the biographies of some still-active officers were not updated and tried to use the latest information, but we did not attempt to independently verify the accuracy or completeness of each document.[3]

The remainder of this section summarizes the results of our evaluation for each of the reserve components and joint staffs throughout DoD. (Appendix H contains detailed results.) We found few cases where the remaining positions serve primarily developmental purposes.

Air National Guard. More than one-third of all positions on the list of 95 belong to the Air National Guard. Twenty-five positions are coded for O-7 and another eight for O-8. The 25 O-7 positions are of two types: Air National Guard assistants to various senior leaders, and Air National Guard staff. Many of these positions are dual-

---

[3] We also tried to accommodate multiple ways that a position could be described. In searching for Air Staff positions we searched for "A1", "A-1", "A 1", "A/1." We searched for concatenations and full titles of commands (for example, "USTRANSCOM," "TRANSCOM," "Transportation Command"). These are just examples of our approach—we attempted permutations of other titles and organizations when appropriate. Still, we cannot aver that our search was completely exhaustive, given the possibilities of unconventional abbreviations and simple typographical errors.

hatted, typically with a state AAG position, and most have had only two or three people in the position. For some of the positions, one person might have gone on to another assignment (less often getting promoted to O-8), but all in all the evidence that these positions are used to develop general officers is weak. This is not to say that the Air National Guard does not use these positions to develop its officers, nor do we say that these positions could not be used in that way. But the empirical evidence is lacking, and arguments for the positions' developmental benefits must be made on some other basis.

Two positions that may be exceptions to this general conclusion are the Air National Guard assistants to the Chief Nurse of the Air Force and to the Chief of Chaplains of the Air Force. These are positions in nonline communities that may have a very narrow pipeline,[4] and both have produced people in their respective communities that went on to more-senior positions as O-8s.

Army National Guard. Among the six Army National Guard positions on the list of 95, one position stands out as possibly being used to develop general officers—the special assistant to the director, Army National Guard. There is sometimes more than one officer in this position, and the current database contains two requirements for this position. Those count for two of the 95 positions we discuss in this section.

NGB Joint Staff. Given the career patterns of officers who have recently held them, none of the seven NGB positions on the list of 95 has compelling empirical evidence that it is used to develop general officers.

Air Force Reserve. Mobilization assistants account for about three-quarters of all Air Force Reserve general officer positions. Many Air Force Reserve general officers serve as mobilization assistants at some point in their careers. Of the nine Air Force Reserve positions on the list of 95, five are mobilization assistants to the commander of a numbered Air Force—the 2nd, 8th, 18th, 19th, and 20th.[5] Of those five, the 2nd, 19th, and 20th mobilization assistants are coded for an O-7. Ten officers have served in one of those three positions and at least five reached the grade of O-8, and several O-7s are still active. Including the two positions coded for an O-8, at least 25 officers have served in one of these numbered Air Force positions, including the current Chief of the Air Force Reserve. Serving as a mobilization assistant to the commander of a numbered Air Force appears to be a common stepping stone to higher grades and responsibilities. A reasonable case could be made that all are used for developmental purposes. Two other positions also could be considered developmental: mobilization assistant

---

[4] There are two basic types of commissioned officers: A *line* officer exercises general command authority and is eligible for operational command positions. A *nonline* officer normally exercises authority within a specialty and includes medical officers, chaplains, and lawyers in the Army and the Navy.

[5] Each numbered Air Force has a mobilization assistant to the commander (14 total, currently). The reason these five positions are on the list of 95 but the others are not comes down to responses to two questions—whether the position has subordinates (the responses for all 14 numbered Air Force mobilization assistants were either *Yes* or *Unknown*) and the frequency of interaction with other senior officials.

to the commander, 618th Air Operations Center (Tanker Airlift Control Center) and mobilization assistant to the commander, Curtis E. Lemay Center for Doctrine Development and Education.

Army Reserve. Given the career patterns of officers who have recently held them, none of the four Army Reserve positions on the list of 95 has compelling empirical evidence that it is used to develop general officers.

Marine Corps Reserve. There are several assistant commandant positions on the list of 95, but all are vacant and there is no record of any incumbents, other than the Assistant Deputy Commandant for Aviation (Mobilization). It does not appear to be used to develop general officers.

Navy Reserve. Given the career patterns of officers who have recently held them, neither of the two Navy Reserve positions on the list of 95 has compelling empirical evidence that it is used to develop flag officers.

Joint Staff. Several Air Force Reserve officers who served as the J-3 mobilization assistants at U.S. Strategic Command and North American Aerospace Defense Command have gone on to additional assignments and promotions, so a case could be made that the Air Force Reserve uses these as developmental opportunities. But the developmental rationale is challenging for joint positions that can be filled by multiple services and reserve components. Whether any joint position is developmental is largely up to the service in which the officer serves, rather than the Joint Staff itself.

Our empirical review of the career trajectories of general and flag officers assigned to selected positions—those on the list of 95—found little evidence that those positions are used to develop general and flag officers. To be clear, we are not claiming that these positions have no developmental merit. Nor are we claiming that the services do not attempt to assign officers to these positions for developmental reasons. But the empirical evidence that they do so is weak. For some of these positions, the issue is simply that very few officers have filled them so far. Arguments for the developmental benefits must be made on philosophical, rather than empirical, grounds. There were a few exceptions to this overall conclusion, noted in the previous discussion.

Findings and Recommendations

Our findings support two broad conclusions with regard to the relationship of development to general and flag officer requirements for the National Guard and reserve.

- First, the services view development as broadening, and they use assignments to achieve that goal. They have a relatively shared understanding of the assignments necessary to develop most general and flag officers for more-senior leadership, including command at the O-7 level, planning and budgeting, and joint experience.

- Second, the current slate of general and flag officer positions is generally adequate to support this process. The most fundamental question for this analysis was whether the services possess a sufficient array of jobs to create the desired number of well-rounded officers. If the services routinely lamented their inability to get officers a certain type of experience, or were worried that their needs were concentrated too heavily in a few areas to generate enough well-rounded leaders, there could be an argument for adjusting the requirements in the name of development, but that is not the case.

Neither our meetings with general and flag officer management offices or senior leaders nor our analysis of career paths suggested any serious gaps in the ability to generate a sufficient number of well-rounded officers for more-senior positions. There appears to be a sufficient range and diversity of positions to provide the necessary experiences. And just as important, constraints that do arise are usually a result of limitations on National Guard and reserve assignment processes, such as the location and availability of officers, not the absence of a particular job.

There are two relatively minor exceptions to this conclusion. One is in regard to joint positions, which were cited in several discussions with at least two services as being in high demand for RC G/FOs and which sometimes posed a particular challenge to achieving developmental goals. But all the services recognized that these are at a premium even for the active component and for officer ranks well below O-7, and even in this case, no service suggested that the shortage of joint general and flag officer positions prevented them from developing a sufficient bench of senior officers. A second exception has to do with educational experiences: All the services noted a significant shortage of seats at CAPSTONE and even at the war colleges before the general and flag officer stage.[6] These experiences are critical to building the relationships necessary for effectiveness at more-senior levels. Arguably the simplest policy change for DoD might be to boost the size or number of CAPSTONE classes to allow 20 to 30 additional newly promoted reserve component O-7s to attend each year.

Development, therefore, does not appear to offer clear or specific guidance to judging National Guard and reserve general and flag officer positions. The process of leader development aims to prepare a sufficient number of officers for selection into senior positions, but it should not drive requirements.

However, two important recommendations did emerge from this analysis:

- **Future reductions in general and flag officer requirements or authorizations should not disproportionately affect those positions that are viewed as devel-**

---

[6] CAPSTONE, a course designed for general and flag officers, was developed to focus on effective planning and employment of U.S. forces in joint and combined operations. Created in 1982, the Goldwater-Nichols DoD Reorganization Act of 1986 subsequently mandated that all newly selected general and flag officers attend CAPSTONE.

opmentally crucial. The services should strive to maintain a good balance of command, joint, headquarters staff, international, combatant command, major command, and other types of positions. The single most common and important route to development for senior responsibilities is by broadening the range of assignments that officers can hold. Cuts that fall disproportionately on less "core" or typical command assignments (such as general and flag officers assigned in the joint environment) could have a serious effect on the developmental potential of general and flag officer requirements.

- **Best practices, ideas, and insights from current leader development efforts in the services should be shared to catalyze further experimentation and innovation in leader development.** While this analysis has been confined to the implications of development for general and flag officer requirements, it has generated broader insights into the current status of officer development programs in the various services. Innovative and promising efforts are under way in many of the reserve components to enhance career-long leader development, both before and after officers are promoted to the general and flag officer ranks.

CHAPTER SIX

Conclusions

Our examination of how the services manage RC G/FO requirements and authorizations revealed a process that generally works, that is conducted within the limitations set in statute, and that provides an adequate number of general and flag officers to meet mission needs and develop a bench of qualified officers from which to fill senior leadership positions. That said, there are opportunities to make the system work more effectively. Absent overarching guidance, the services approach general officer management in different ways, driven by their individual management philosophies. These distinctions can result in differences across services and across similar positions that are often difficult to explain. Our report closes with a summary of the broad conclusions that emerged from our research and recommendations for how the system can be improved in the future.

Current System of Authorized Strength

The services report that RC G/FO authorized strength is sufficient for the effective management and leadership of the reserve components. The services report that the current system of basic authorized strength and exemptions provides an adequate number of RC G/FOs to fulfill the varied roles in which these officers serve. The exemptions, in particular, are necessary to meet the full range of RC G/FO requirements. While there may be particular requirements that the management offices are unable to fill from time to time, there is no consistent type or set of requirements that they would like to fill but for which they do not have the authorized strength.

Exemptions to the basic authorized strength are appropriate in both structure and number. The general and flag officer management offices commented on the complexity of the many exemptions and suggested that a simpler system combining exemptions and the basic authorized strength into a single pool would be more understandable and easier to manage, and could help mitigate the perception that there are *too many* RC G/FOs. We analyzed the pros and cons of such a model and could not find sufficient evidence to support changing the current system. Eliminating exemptions would reduce needed flexibility in managing RC G/FOs because the services would lose the ability to explicitly designate authorizations for particular roles that need to be filled—and would be counter to the very purpose of these exemptions.

Reserve Component General and Flag Officer Requirements Process

The services' current requirements and authorizations processes work but would benefit from a systematic, comprehensive review process. Although each of the services has a process for generating and validating RC G/FO requirements and allocating authorizations to these positions, the approaches vary and are generally ad hoc. Only the Air Force Reserve reported having a transparent, comprehensive, well-documented method for validating and prioritizing its RC G/FO requirements. The services' general and flag officer management offices report that they do consider a number of characteristics for validating and revalidating positions to determine if the position warrants a general or flag officer; however, the single most important consideration is military judgment of service and joint senior leaders. All of the services report that they are able to meet their needs and critical needs are not going unfilled.

Though these processes generally work, they do have the potential to result in inconsistencies across the services and over time. This is particularly the case when differences arise regarding individual service philosophies on the appropriate roles for RC G/FOs. These differences can, for instance, result in positions with similar responsibilities being filled by individuals of different rank.

All previous general and flag officer requirements studies we reviewed (for the active and reserve components) emphasize the difficulties in determining how positions requiring general and flag officers are designated or how many positions should be authorized. We recognize that military senior leaders need to have the latitude to make these decisions as they organize, train, equip, and employ forces, but general guidelines covering all the services would help in determining which positions require general and flag officers—and justifying these to Congress.

We recommend that OUSD(P&R) release suggested guidelines for evaluating and validating positions that should be designated for RC G/FOs based on their expected roles and responsibilities. These guidelines would encourage consistency across service and joint positions and would better posture the department for future RC G/FO reviews. We encourage the services to develop a formal, visible prioritization process for creating and validating RC G/FO positions, such as the one currently being used by the Air Force Reserve, and to document the process and their intent. The methodology developed in Chapter Four can provide the basis for developing these guidelines and a way to validate or revalidate general and flag officer positions being considered.

The number of RC G/FO requirements exceeds the authorized strength. Each service RC G/FO management office reports that there are slightly more requirements designated for RC G/FOs than can be filled within the limits on authorized strength.

That notwithstanding, each of the services intentionally allows some room between the number of RC G/FO positions filled and its authorized strength for necessary practical management issues that arise as a result of officer promotions and retirements, and also to ensure that authorizations are available if needed for emerging requirements.

As of February 2015, 66 validated requirements for RC G/FOs were vacant because of either limitations on RC G/FO authorized strength or lack of funding. These vacancies raise the question of whether critical needs are going unfilled. The services suggest that the answer is no—they report that authorized strength limitations do not interfere with accomplishing the mission. Instead, positions are prioritized, available authorizations are allocated to positions according to critical need, and the vacant positions do not generate significant risk in going unfilled.

The services' methods for ensuring compliance with RC G/FO authorized strength are adequate but would benefit from consistent cross-DoD tracking. The general and flag officer management offices expend considerable time and attention ensuring that they are adhering to the complex set of statutory limitations to general and flag officer authorized strength. Each of the services' management offices has its own means for tracking assignments. We reviewed these tracking systems and found that they are accurately, appropriately, and consistently ensuring compliance with statutory limits on the number of RC G/FOs.

It took significant time and coordination between the RAND research team and the general and flag officer management offices to understand each service's management approach and tracking method and, in turn, validate that it was properly interpreting and adhering to statutory limitations. The task was challenging because of inconsistency in terminology, methods of accounting, and depictions of the categories of authorizations and exemptions across the services. We recommend that OUSD(P&R) institute a single system for tracking RC G/FO authorized strength, both under the basic authorized strength and for the individual categories of exemptions, that would use a common taxonomy and provide a DoD-wide status of RC G/FO authorizations. With the services using a common tracking approach, OUSD(P&R) can better ensure that the number of RC G/FOs does not exceed statutory limitation and can track the use of exemptions over time more efficiently.

Authorized strength should not be driven by changes in total end strength or the use of RC G/FOs in times of war or national emergency. We reviewed two factors that have the potential to affect the authorized strength for RC G/FOs—end strength numbers and the use of RC G/FOs in times of war or national emergency. We found that the size of the RC G/FO corps should not be tied to either factor. In the case of end strength, basing general and flag officer authorized strength on reserve component end strength could limit the use of RC G/FOs where they provide expertise, gain experience, and fill in for skill shortages in the active component, but do not lead large numbers of troops. As such, limiting RC G/FO authorized strength based on end strength would be counterproductive.

Opportunities for Efficiencies

There are potential opportunities to gain efficiencies by eliminating or downgrading a number of RC G/FO requirements. Our assessment identified only 27 of 634 positions (about 4 percent) as candidates for further review by the services to either revalidate them as RC G/FO requirements or consider them for elimination or downgrading. Important contextual, qualitative, and subjective factors make it impossible for us to make purely analytical determinations on these 27 positions. Some stand out fairly clearly as reasonable positions for general or flag officers, and service review may indeed revalidate all or most of these. A second, more stringent filtering produced 95 positions, but review of some additional factors suggests that about half of these (which include the original 27) might be worthy of detailed review for possible reductions.

Our empirical review of the career trajectories of general and flag officers assigned to these 95 positions found little evidence that requirements can be fully justified on the basis of developmental opportunities. We are not claiming that these positions have no developmental merit; neither are we claiming that the services do not attempt to assign officers to these positions for developmental reasons. But our empirical review of officer career paths produced little evidence that they do. For some of these positions, the issue is simply that very few officers have filled them so far, in part because they are new requirements or newly authorized positions. Thus, arguments for developmental benefits must be made on philosophical, rather than empirical, grounds.

Positions described as mobilization or special assistants or National Guard assistants (most of which are filled by IMAs) are the requirements category that raises the most questions. These positions have loosely defined peacetime roles that can be described as liaisons between their commands and the reserve components. Their roles upon mobilization are not well articulated. As these constitute one-third of the RC G/FO requirements, not counting those requirements for state TAGs and AAGs, it would help justify these positions and assuage congressional concerns if there were overarching service policies and standards for mobilization or National Guard assistants, addressing their expected roles both in peacetime and during mobilization.

OUSD(P&R) should assist the services in establishing overarching guidance on how IMAs should be used and the roles they should play. While flexibility to tailor assignments is essential, there should be a common foundation that establishes the basis for how all the services employ RC G/FO IMAs.

There are few opportunities to convert RC G/FO positions to senior-level civilian positions. The stakeholders we talked to did not see an opportunity for converting RC G/FO positions to senior civilian positions. In some positions, the status of a military general or flag officer was necessary for effective interactions with other senior military officials and stakeholders. For positions with specific military command requirements, such as deputy or vice commanders, an SES position would not be appropriate. The biggest impediment is that many of the RC G/FO positions are part-time, although the value of the position for development of RC G/FOs may also be an important consideration.

RC G/FO requirements that have emerged since 9/11 broadly reflect the needs of the past 15 years, but should be reviewed as operational requirements change. Many RC G/FO requirements added since 9/11 are closely related to enduring changes in force structure and organizations, but others are more closely tied to operational requirements that continue to evolve. We recommend that the services regularly review the positions more closely tied to operational requirements, particularly in the Central Command area of responsibility, to ensure that limited general and flag officer resources are dedicated to the highest-priority missions. As operations end or evolve, there may be opportunities to free up valuable RC G/FO resources for use elsewhere. The reduction in funding for overseas contingencies provides a natural opportunity to review the status of the general and flag officer requirements currently being funded with these resources.

Considering Officer Development in Setting Requirements

The services possess a sufficient array of positions to create the desired number of well-rounded officers. The services view development as broadening and use assignments to achieve that goal. They have a relatively shared understanding of the assignments necessary to develop most RC G/FOs for more-senior leadership, including command at the O-7 level, planning and budgeting, and joint experience. The current slate of RC G/FO requirements is generally adequate to support this process. None of the offices or senior leaders with whom we met, nor our analysis of career paths, suggested any serious gaps in the ability to generate a sufficient number of well-rounded officers for more-senior positions. There appears to be a sufficient range and diversity of positions to provide the necessary experiences. And just as important, where constraints arise, they are usually a result of limitations on National Guard and reserve assignment processes, such as the location and availability of officers, and not because a particular job opportunity does not exist.

Though our analysis focused on whether development should be a factor in establishing requirements for RC G/FOs, we also gained broader insights into the current status of officer development programs in the various services. Innovative and promising efforts are under way in many of the reserve components to enhance career-long leader development, both before and after officers are promoted to the general and flag officer ranks. The services should share best practices, ideas, and insights from current leader development efforts to catalyze further experimentation and innovation in leader development.

Development considerations offer only general guidance when judging RC G/FO requirements. The process of leader development aims to prepare a sufficient number of officers to provide appropriate selectivity for senior positions, but it should not drive requirements. Future reductions in RC G/FO positions should not disproportionately affect those positions that are viewed as developmentally crucial. The services should strive to maintain a good balance of command, joint, headquarters staff, international, combatant command, major command, and other types of positions. The single most common and important route to development for senior responsibilities is by broadening the range of assignments that officers can hold. Cuts that fall disproportionately on less "core" or typical command assignments (such as general and flag officers assigned in the joint environment) could have a serious effect on the developmental potential of RC G/FO requirements. The value of a position from a developmental standpoint may not always be apparent from some of the other criteria used to judge whether a position requires a general officer.

APPENDIX A

Key Statutory Provisions

Chapter Two described the statutory authority applicable to general and flag officers in the active and reserve component. Table A.1 contains excerpts from the key statutory provisions contained in Title 10 U.S. Code.

Table A.1
Provision Excerpts

Section	Quoted Content from 10 U.S.C.
10 U.S. Code Section 525	Distribution of commissioned officers on active duty in general officer and flag officer grades
	(a) For purposes of the applicable limitation in section 526(a) of this title [10 USCS § 526(a)] on general and flag officers on active duty, no appointment of an officer on the active duty list may be made as follows: (1) in the Army, if that appointment would result in more than— (A) 7 officers in the grade of general; (B) 46 officers in a grade above the grade of major general; or (C) 90 officers in the grade of major general; (2) in the Air Force, if that appointment would result in more than— (A) 9 officers in the grade of general; (B) 44 officers in a grade above the grade of major general; or (C) 73 officers in the grade of major general; (3) in the Navy, if that appointment would result in more than— (A) 6 officers in the grade of admiral; (B) 33 officers in a grade above the grade of rear admiral; or (C) 50 officers in the grade of rear admiral; (4) in the Marine Corps, if that appointment would result in more than— (A) 2 officers in the grade of general; (B) 15 officers in a grade above the grade of major general; or (C) 23 officers in the grade of major general. (b) The limitations of subsection (a) do not include the following: (1) An officer released from a joint duty assignment, but only during the 60-day period beginning on the date the officer departs the joint duty assignment, except that the Secretary of Defense may authorize the Secretary of a military department to extend the 60-day period by an additional 120 days, but no more than three officers from each armed forces may be on active duty who are excluded under this paragraph. (2) The number of officers required to serve in joint duty assignments as authorized by the Secretary of Defense under section 526(b) [10 USCS § 526(b)] for each military service.

Table A.1—Continued

Section	Quoted Content from 10 U.S.C.
	(c) (1) Subject to paragraph (3), the President— (A) may make appointments in the Army, Air Force, and Marine Corps in the grades of lieutenant general and general in excess of the applicable numbers determined under this section if each such appointment is made in conjunction with an offsetting reduction under paragraph (2); and (B) may make appointments in the Navy in the grades of vice admiral and admiral in excess of the applicable numbers determined under this section if each such appointment is made in conjunction with an offsetting reduction under paragraph (2). (2) For each appointment made under the authority of paragraph (1) in the Army, Air Force, or Marine Corps in the grade of lieutenant general or general or in the Navy in the grade of vice admiral or admiral, the number of appointments that may be made in the equivalent grade in one of the other armed forces (other than the Coast Guard) shall be reduced by one. When such an appointment is made, the President shall specify the armed force in which the reduction required by this paragraph is to be made. (3) (A) The number of officers that may be serving on active duty in the grades of lieutenant general and vice admiral by reason of appointments made under the authority of paragraph (1) may not exceed 15. (B) The number of officers that may be serving on active duty in the grades of general and admiral by reason of appointments made under the authority of paragraph (1) may not exceed 5. (4) Upon the termination of the appointment of an officer in the grade of lieutenant general or vice admiral or general or admiral that was made in connection with an increase under paragraph (1) in the number of officers that may be serving on active duty in that armed force in that grade, the reduction made under paragraph (2) in the number of appointments permitted in such grade in another armed force by reason of that increase shall no longer be in effect.
	(d) An officer continuing to hold the grade of general or admiral under section 601(b)(5) of this title [10 USCS § 601(b)(5)] after relief from the position of Chairman of the Joint Chiefs of Staff, Chief of Staff of the Army, Chief of Naval Operations, Chief of Staff of the Air Force, or Commandant of the Marine Corps shall not be counted for purposes of this section.
	(e) The following officers shall not be counted for purposes of this section: (1) An officer of that armed force in the grade of brigadier general or above or, in the case of the Navy, in the grade of rear admiral (lower half) or above, who is on leave pending the retirement, separation, or release of that officer from active duty, but only during the 60-day period beginning on the date of the commencement of such leave of such officer. (2) At the discretion of the Secretary of Defense, an officer of that armed force who has been relieved from a position designated under section 601(a) of this title [10 USCS § 601(a)] or by law to carry one of the grades specified in such section, but only during the 60-day period beginning on the date on which the assignment of the officer to the first position is terminated or until the officer is assigned to a second such position, whichever occurs first.
	(f) An officer while serving as Attending Physician to the Congress is in addition to the number that would otherwise be permitted for that officer's armed force for officers serving on active duty in grades above brigadier general or rear admiral (lower half) under subsection (a).

Table A.1—Continued

Section	Quoted Content from 10 U.S.C.
	(g) (1) The limitations of this section do not apply to a reserve component general or flag officer who is on active duty for a period in excess of 365 days, but not to exceed three years, except that the number of officers from each reserve component who are covered by this subsection and are not serving in a position that is a joint duty assignment for purposes of chapter 38 of this title [10 USCS §§ 661 et seq.] may not exceed 5 per component, unless authorized by the Secretary of Defense. (2) Not later than 30 days after authorizing a number of reserve component general or flag officers in excess of the number specified in paragraph (1), the Secretary of Defense shall notify the Committees on Armed Services of the Senate and the House of Representatives of such authorization, and shall include with such notice a statement of the reason for such authorization.
10 U.S. Code **Section** 526	Authorized strength: general and flag officers on active duty
	(a) Limitations. The number of general officers on active duty in the Army, Air Force, and Marine Corps, and the number of flag officers on active duty in the Navy, may not exceed the number specified for the armed force concerned as follows: (1) For the Army, 231. (2) For the Navy, 162. (3) For the Air Force, 198. (4) For the Marine Corps, 61. (b) Limited exclusion for joint duty requirements. (1) The Secretary of Defense may designate up to 310 general officer and flag officer positions that are joint duty assignments for purposes of chapter 38 of this title [10 USCS §§ 661 et seq.] for exclusion from the limitations in subsection (a). The Secretary of Defense shall allocate those exclusions to the armed forces based on the number of general or flag officers required from each armed force for assignment to these designated positions. (2) Unless the Secretary of Defense determines that a lower number is in the best interest of the Department, the minimum number of officers serving in positions designated under paragraph (1) for each armed force shall be as follows: (A) For the Army, 85. (B) For the Navy, 61. (C) For the Air Force, 73. (D) For the Marine Corps, 21. (3) The number excluded under paragraph (1) and serving in positions designated under that paragraph-- (A) in the grade of general or admiral may not exceed 20; (B) in a grade above the grade of major general or rear admiral may not exceed 68; and (C) in the grade of major general or rear admiral may not exceed 144. (4) Not later than 30 days after determining to raise or lower a number specified in paragraph (2), the Secretary of Defense shall notify the Committees on Armed Services of the Senate and the House of Representatives of such determination.

Table A.1—Continued

Section	Quoted Content from 10 U.S.C.
	(5) (A) The Chairman of the Joint Chiefs of Staff may designate up to 15 general and flag officer positions in the unified and specified combatant commands, and up to three general and flag officer positions on the Joint Staff, as positions to be held only by reserve component officers who are in a general or flag officer grade below lieutenant general or vice admiral. Each position so designated shall be considered to be a joint duty assignment position for purposes of chapter 38 of this title [10 USCS §§ 661 et seq.]. (B) A reserve component officer serving in a position designated under subparagraph (A) while on active duty under a call or order to active duty that does not specify a period of 180 days or less shall not be counted for the purposes of the limitations under subsection (a) and under section 525 of this title [10 USCS § 525] if the officer was selected for service in that position in accordance with the procedures specified in subparagraph (C). (C) Whenever a vacancy occurs, or is anticipated to occur, in a position designated under subparagraph (A)-- (i) the Secretary of Defense shall require the Secretary of the Army to submit the name of at least one Army reserve component officer, the Secretary of the Navy to submit the name of at least one Navy Reserve officer and the name of at least one Marine Corps Reserve officer, and the Secretary of the Air Force to submit the name of at least one Air Force reserve component officer for consideration by the Secretary for assignment to that position; and (ii) the Chairman of the Joint Chiefs of Staff may submit to the Secretary of Defense the name of one or more officers (in addition to the officers whose names are submitted pursuant to clause (i)) for consideration by the Secretary for assignment to that position. (D) Whenever the Secretaries of the military departments are required to submit the names of officers under subparagraph (C)(i), the Chairman of the Joint Chiefs of Staff shall submit to the Secretary of Defense the Chairman's evaluation of the performance of each officer whose name is submitted under that subparagraph (and of any officer whose name the Chairman submits to the Secretary under subparagraph (C)(ii) for consideration for the same vacancy). (E) Subparagraph (B) does not apply in the case of an officer serving in a position designated under subparagraph (A) if the Secretary of Defense, when considering officers for assignment to fill the vacancy in that position which was filled by that officer, did not have a recommendation for that assignment from each Secretary of a military department who (pursuant to subparagraph (C)) was required to make such a recommendation.

Table A.1—Continued

Section	Quoted Content from 10 U.S.C.
	(c) Exclusion of certain reserve officers. (1) The limitations of this section do not apply to a reserve component general or flag officer who is on active duty for training or who is on active duty under a call or order specifying a period of less than 180 days. (2) The limitations of this section also do not apply to a number, as specified by the Secretary of the military department concerned, of reserve component general or flag officers authorized to serve on active duty for a period of not more than 365 days. The number so specified for an armed force may not exceed the number equal to 10 percent of the authorized number of general or flag officers, as the case may be, of that armed force under section 12004 of this title [10 USCS § 12004]. In determining such number, any fraction shall be rounded down to the next whole number, except that such number shall be at least one. (3) The limitations of this section do not apply to a reserve component general or flag officer who is on active duty for a period in excess of 365 days but not to exceed three years, except that the number of such officers from each reserve component who are covered by this paragraph and not serving in a position that is a joint duty assignment for purposes of chapter 38 of this title [10 USCS §§ 661 et seq.] may not exceed 5 per component, unless authorized by the Secretary of Defense. (d) Exclusion of certain officers pending separation or retirement or between senior positions. The limitations of this section do not apply to a general or flag officer who is covered by an exclusion under section 525(e) of this title [10 USCS § 525(e)]. (e) Exclusion of Attending Physician to the Congress. The limitations of this section do not apply to the general or flag officer who is serving as Attending Physician to the Congress. (f) Temporary exclusion for assignment to certain temporary billets. (1) The limitations in subsection (a) and in section 525(a) of this title [10 USCS § 525(a)] do not apply to a general or flag officer assigned to a temporary joint duty assignment designated by the Secretary of Defense. (2) A general or flag officer assigned to a temporary joint duty assignment as described in paragraph (1) may not be excluded under this subsection from the limitations in subsection (a) for a period of longer than one year. (g) Exclusion of officers departing from joint duty assignments. The limitations in subsection (a) do not apply to an officer released from a joint duty assignment, but only during the 60-day period beginning on the date the officer departs the joint duty assignment. The Secretary of Defense may authorize the Secretary of a military department to extend the 60-day period by an additional 120 days, except that not more than three officers on active duty from each armed force may be covered by an extension under this sentence at the same time.

Table A.1—Continued

Section	Quoted Content from 10 U.S.C.
	(h) Active-duty baseline. (1) Notice and wait requirement. If the Secretary of a military department proposes an action that would increase above the baseline the number of general officers or flag officers of an armed force under the jurisdiction of that Secretary who would be on active duty and would count against the statutory limit applicable to that armed force under subsection (a), the action shall not take effect until after the end of the 60-calendar day period beginning on the date on which the Secretary provides notice of the proposed action, including the rationale for the action, to the Committees on Armed Services of the House of Representatives and the Senate. (2) Baseline defined. For purposes of paragraph (1), the term «baseline» for an armed force means the lower of— (A) the statutory limit of general officers or flag officers of that armed force under subsection (a); or (B) the actual number of general officers or flag officers of that armed force who, as of January 1, 2014, counted toward the statutory limit of general officers or flag officers of that armed force under subsection (a). (3) Limitation. If, at any time, the actual number of general officers or flag officers of an armed force who count toward the statutory limit of general officers or flag officers of that armed force under subsection (a) exceeds such statutory limit, then no increase described in paragraph (1) for that armed force may occur until the general officer or flag officer total for that armed force is reduced below such statutory limit.
	(i) Joint duty assignment baseline. (1) Notice and wait requirement. If the Secretary of Defense, the Secretary of a military department, or the Chairman of the Joint Chiefs of Staff proposes an action that would increase above the baseline the number of general officers and flag officers of the armed forces in joint duty assignments who count against the statutory limit under subsection (b)(1), the action shall not take effect until after the end of the 60-calendar day period beginning on the date on which the Secretary or Chairman, as the case may be, provides notice of the proposed action, including the rationale for the action, to the Committees on Armed Services of the House of Representatives and the Senate. (2) Baseline defined. For purposes of paragraph (1), the term «baseline» means the lower of— (A) the statutory limit on general officer and flag officer positions that are joint duty assignments under subsection (b)(1); or (B) the actual number of general officers and flag officers who, as of January 1, 2014, were in joint duty assignments counted toward the statutory limit under subsection (b)(1). (3) Limitation. If, at any time, the actual number of general officers and flag officers in joint duty assignments counted toward the statutory limit under subsection (b)(1) exceeds such statutory limit, then no increase described in paragraph (1) may occur until the number of general officers and flag officers in joint duty assignments is reduced below such statutory limit.
	(j) Annual report on general officer and flag officer numbers. Not later than March 1, 2015, and each March 1 thereafter, the Secretary of Defense shall submit to the Committees on Armed Services of the House of Representatives and the Senate a report specifying— (1) the numbers of general officers and flag officers who, as of January 1 of the calendar year in which the report is submitted, counted toward the service-specific limits of subsection (a); and (2) the number of general officers and flag officers in joint duty assignments who, as of such January 1, counted toward the statutory limit under subsection (b)(1).

Table A.1—Continued

Section	Quoted Content from 10 U.S.C.
10 U.S. Code **Section** 527	Authority to suspend sections 523, 525, and 526
	In time of war, or of national emergency declared by Congress or the President after November 30, 1980, the President may suspend the operation of any provision of section 523, 525, or 526 of this title [10 USCS § 523, 525, or 526]. So long as such war or national emergency continues, any such suspension may be extended by the President. Any such suspension shall, if not sooner ended, end on the last day of the two-year period beginning on the date on which the suspension (or the last extension thereof) takes effect or on the last day of the one-year period beginning on the date of the termination of the war or national emergency, whichever occurs first. With respect to the end of any such suspension, the preceding sentence supersedes the provisions of title II of the National Emergencies Act (50 U.S.C. 1621-1622) which provide that powers or authorities exercised by reason of a national emergency shall cease to be exercised after the date of the termination of the emergency.
10 U.S. Code **Section** 528	Officers serving in certain intelligence positions: military status; application of distribution and strength limitations; pay and allowances
	(a) Military status. An officer of the armed forces, while serving in a position covered by this section— (1) shall not be subject to supervision or control by the Secretary of Defense or any other officer or employee of the Department of Defense, except as directed by the Secretary of Defense concerning reassignment from such position; and (2) may not exercise, by reason of the officer's status as an officer, any supervision or control with respect to any of the military or civilian personnel of the Department of Defense except as otherwise authorized by law. (b) Director and Deputy Director of CIA. When the position of Director or Deputy Director of the Central Intelligence Agency is held by an officer of the armed forces, the position, so long as the officer serves in the position, shall be designated, pursuant to subsection (b) of section 526 of this title [10 USCS § 526], as one of the general officer and flag officer positions to be excluded from the limitations in subsection (a) of such section. (c) Associate Director of Military Affairs, CIA. When the position of Associate Director of Military Affairs, Central Intelligence Agency, or any successor position, is held by an officer of the armed forces, the position, so long as the officer serves in the position, shall be designated, pursuant to subsection (b) of section 526 of this title [10 USCS § 526], as one of the general officer and flag officer positions to be excluded from the limitations in subsection (a) of such section. (d) Officers serving in Office of DNI. When a position in the Office of the Director of National Intelligence designated by agreement between the Secretary of Defense and the Director of National Intelligence is held by a general officer or flag officer of the armed forces, the position, so long as the officer serves in the position, shall be designated, pursuant to subsection (b) of section 526 of this title [10 USCS § 526], as one of the general officer and flag officer positions to be excluded from the limitations in subsection (a) of such section. However, not more than five of such positions may be included among the excluded positions at any time. (e) Effect of appointment. Except as provided in subsection (a), the appointment or assignment of an officer of the armed forces to a position covered by this section shall not affect— (1) the status, position, rank, or grade of such officer in the armed forces; or (2) any emolument, perquisite, right, privilege, or benefit incident to or arising out of such status, position, rank, or grade.

Table A.1—Continued

Section	Quoted Content from 10 U.S.C.
	(f) Military pay and allowances. (1) An officer of the armed forces on active duty who is appointed or assigned to a position covered by this section shall, while serving in such position and while remaining on active duty, continue to receive military pay and allowances and shall not receive the pay prescribed for such position. (2) Funds from which pay and allowances under paragraph (1) are paid to an officer while so serving shall be reimbursed as follows: (A) For an officer serving in a position within the Central Intelligence Agency, such reimbursement shall be made from funds available to the Director of the Central Intelligence Agency. (B) For an officer serving in a position within the Office of the Director of National Intelligence, such reimbursement shall be made from funds available to the Director of National Intelligence. (g) Covered positions. The positions covered by this section are the positions specified in subsections (b) and (c) and the positions designated under subsection (d).
10 U.S. Code **Section** 12001	Authorized strengths: reserve components
	(a) Whenever the authorized strength of a reserve component (other than the Coast Guard Reserve) is not prescribed by law, it shall be prescribed by the President. (b) Subject to the authorized strength of the reserve component concerned, the authorized strength of each reserve component (other than the Coast Guard Reserve) in members in each grade is that which the Secretary concerned determines to be necessary to provide for mobilization requirements. The Secretary shall review these determinations at least once each year and revise them if he considers it necessary. However, a member of the reserve component concerned may not, as a result of such a determination, be reduced in the member's reserve grade without the member's consent.
10 U.S. Code **Section** 12002	Authorized strengths: Army and Air Force reserve components, exclusive of members on active duty
	(a) The authorized strengths of the National Guard and the reserve components of the Army and the Air Force, exclusive of members who are included in the strengths authorized for members of the Army and Air Force, respectively, on active duty, are as follows: Army National Guard and the Army National Guard of the United States...600,000 Army Reserve...980,000 Air National Guard and the Air National Guard of the United States...150,000 Air Force Reserve...500,000 (b) The strength authorized by this section for the Army National Guard and the Army National Guard of the United States, and the strength authorized by this section for the Air National Guard and the Air National Guard of the United States, shall be allocated among the States.
10 U.S. Code **Section** 12003	Authorized strengths: commissioned officers in an active status
	(a) The authorized strengths of the Army, Navy, Air Force, and Marine Corps in reserve commissioned officers, other than commissioned warrant officers and officers on an active-duty list, in an active status are as follows: Army...275,000 Air Force...200,000 Navy...150,000 Marine Corps...24,500

Table A.1—Continued

Section	Quoted Content from 10 U.S.C.
	(b) The authorized strengths prescribed by subsection (a) may not be exceeded unless— (1) the Secretary concerned determines that a greater number is necessary for planned mobilization requirements; or (2) the excess results directly from the operation of a nondiscretionary provision of law.
10 U.S. Code **Section** 12004	Strength in grade: reserve general and flag officers in an active status
	(a) The authorized strengths of the Army, Air Force, and Marine Corps in reserve general officers in an active status, and the authorized strength of the Navy in reserve officers in the grades of rear admiral (lower half) and rear admiral in an active status, are as follows: Army....207 Air Force....157 Navy.... 48 Marine Corps.... 10 (b) The following Army and Air Force reserve officers shall not be counted for purposes of this section: (1) Those serving as adjutants general or assistant adjutants general of a State. (2) Those serving in the National Guard Board. (3) Those counted under section 526 of this title [10 USCS § 526]. (4) Those serving in a joint duty assignment for purposes of chapter 38 of this title [10 USCS §§ 661 et seq.], except that the number of officers who may be excluded under this paragraph may not exceed the number equal to 20 percent of the number of officers authorized for the armed force concerned by subsection (a). (c) (1) The following Navy reserve officers shall not be counted for purposes of this section: (A) Those counted under section 526 of this title [10 USCS § 526]. (B) Those serving in a joint duty assignment for purposes of chapter 38 of this title [10 USCS §§ 661 et seq.], except that the number of officers who may be excluded under this paragraph may not exceed the number equal to 20 percent of the number of officers authorized for the Navy in subsection (a). (2) Not more than 50 percent of the officers in an active status authorized under this section for the Navy may serve in the grade of rear admiral. (3) [Deleted] (4) [Redesignated] (5) [Deleted] (d) The following Marine Corps reserve officers shall not be counted for purposes of this section: (1) Those counted under section 526 of this title [10 USCS § 526]. (2) Those serving in a joint duty assignment for purposes of chapter 38 of this title [10 USCS §§ 661 et seq.], except that the number of officers who may be excluded under this paragraph may not exceed the number equal to 20 percent of the number of officers authorized for the Marine Corps in subsection (a). (e)(1) A reserve general officer of the Army or Air Force may not be reduced in grade because of a reduction in the number of general officers authorized under subsection (a). (2) An officer of the Navy Reserve or the Marine Corps Reserve may not be reduced in permanent grade because of a reduction in the number authorized by this section for his grade.

Table A.1—Continued

Section	Quoted Content from 10 U.S.C.
	(f) The limitations in subsection (a) do not apply to an officer released from a joint duty assignment or other non-joint active duty assignment, but only during the 60-day period beginning on the date the officer departs the joint duty or other active duty assignment. The Secretary of Defense may authorize the Secretary of a military department to extend the 60-day period by an additional 120 days, except that not more than three officers in an active status from each reserve component may be covered by an extension under this sentence at the same time.
10 U.S. Code **Section** 12006	Strength limitations: authority to waive in time of war or national emergency
	(f) The limitations in subsection (a) do not apply to an officer released from a joint duty assignment or other non-joint active duty assignment, but only during the 60-day period beginning on the date the officer departs the joint duty or other active duty assignment. The Secretary of Defense may authorize the Secretary of a military department to extend the 60-day period by an additional 120 days, except that not more than three officers in an active status from each reserve component may be covered by an extension under this sentence at the same time.
10 U.S. Code **Sections** 3038; 5143; 5144; 8038	Reserve Chiefs Counted on Active Component Active Duty List
	An officer on active duty for service as the Chief of Army Reserve shall be counted for purposes of the grade limitations under sections 525 and 526 of this title; An officer on active duty for service as the Chief of Navy Reserve shall be counted for purposes of the grade limitations under sections 525 and 526 of this title; An officer on active duty for service as the Commander, Marine Forces Reserve, shall be counted for purposes of the grade limitations under sections 525 and 526 of this title; An officer on active duty for service as the Chief of Air Force Reserve shall be counted for purposes of the grade limitations under sections 525 and 526 of this title.

APPENDIX B

Subject-Matter Expert Interview Participants and Protocol

Subject-Matter Expert Participants

RAND researchers met with several types of organizations to gather information for this research. These included the general and flag officer management offices for both reserve and active components of each service branch and the National Guard, DoD-level organizations that manage joint general and flag officer positions and provide oversight of the general and flag officer placement process, general and flag officer management offices at selected combatant command headquarters, and other organizations with information relevant to the analysis of this topic. Table B.1 lists the interview participants.

In two cases, for the Army and Air Force, RAND researchers met with the management offices for both the reserve component and active component together. The Marine Corps has one office that manages both active and reserve component general and flag officers.

The information provided by the reserve component management offices was key because those offices oversee the day-to-day management of general and flag officers, are responsible for establishing new positions, and interact frequently with Chiefs of the Services and Chiefs of the Reserve to understand their priorities. These offices are in the best position to understand the trends in the philosophies behind the prioritization of general and flag officer requirements and how positions and people are tracked in regard to authorized strength and exemptions. The active component general and flag officer management offices provided information on the conditions by which RC G/FOs are used in active duty positions and in joint positions.

RAND researchers also interviewed the Joint Staff general and flag officer management office responsible for staffing all general and flag officer positions (both active and reserve components) on the Joint Staff and on the staffs of the combatant commands and defense agencies. The Joint Staff general and flag officer management office provided information on the requirements for joint positions and the "Blue Book," a listing of all current joint general and flag officer positions and the current incumbents. We also collected information from the OUSD(P&R) OEPM office, which is responsible for ensuring "that the DoD and the services' personnel management policies, as well as existing and proposed statute, facilitate the management and retention

Table B.1
Interview Participants

Organization	Office Affiliation
Active Army	Department of the Army General Officer Management Office
Army Reserve	General Officer Management Office, Office of the Chief of the Army Reserve
Army National Guard	General Officer Management Office, NGB
Active Navy	Navy Flag Officer Matters
Navy Reserve	Reserve Flag Matters, Office of the Chief of Navy Reserve
Marine Corps Active and Reserve	Senior Leadership Management, Headquarters Marine Corps
Active Air Force	Air Force General Officer Management Office
Air Force Reserve	Air Force Reserve General Officer Management Office
Air National Guard	General Officer Management Office, NGB
Joint and Combatant Commands	General/Flag Officer Management, Joint Staff
Office of the Under Secretary of Defense for Personnel and Readiness	Senior Officer Management, Officer and Enlisted Personnel Management
U.S. Central Command	General/Flag Officer Management, Director of Manpower and Personnel
U.S. Northern Command	General/Flag Officer Management, Director of Manpower and Personnel
U.S. Cyber Command	General/Flag Officer Management, Director of Manpower and Personnel
Reserve Forces Policy Board	Reserve Forces Policy Board Military Executive and Staff
Senior Executive Service	Civilian Senior Executive Management Division, Defense Civilian Personnel Advisory Service

of military personnel in the grades and skills required for personnel and readiness and effective mission accomplishment" (OUSD[P&R], undated). OEPM provided a general overview of the DoD general and flag officer management process and historical information on past studies related to general and flag officer management.

We also contacted offices responsible for general and flag officer management at a sample of combatant command headquarters (U.S. Central Command, U.S. Southern Command, and U.S. Cyber Command). These organizations are "users" of RC G/FOs. Their primary concern is that qualified and experienced general and flag officers are available to fill their full-time and part-time requirements. We also collected information from the Civilian Senior Executive Management Division of the Defense Civilian Personnel Advisory Service to understand the requirements for members of the SES and the characteristics of SES positions that differentiate them from general and flag officer requirements. Finally, we contacted the Reserve Forces Policy Board, the Secretary of Defense's independent advisor providing advice and recommendations on strategies, policies, and practices designed to improve and enhance the capabilities, efficiency, and effectiveness of the reserve components. The board provided reserve component–specific concerns on how the reserve forces might view the roles and attributes for RC G/FOs.

Interview Protocol

Interviews were semistructured and covered the following topics:

- processes to create, validate, and fund RC G/FO requirements
- determination of the sufficiency of RC G/FO authorized strength, including exemptions
- need for requirements to support the development of RC G/FOs
- types of and procedures for developing RC G/FOs
- conditions under which RC G/FO positions might be downgraded to colonel positions or converted to SES positions
- how decisions are made on which and how many general or flag officer requirements to fill when requirements are greater than authorizations
- methods used to track RC G/FO authorizations and compliance with statutes
- conditions under which RC G/FOs are used for active component general and flag officer positions
- use or mobilization of RC G/FOs in times of war or national emergency.

We considered the discussions exploratory and therefore probed with additional related questions as relevant. We have included the interview protocol that was used in our interviews and provided to the offices prior to the meeting.

Establishing, Validating, and Reviewing Requirements and Positions for RC G/FOs in an Active Status

- What is the process for establishing an RC G/FO position?
  - Are the process steps documented?
  - At what level are the positions approved?
- What are the organizational/span-of-control/experience standards applied for determining the grade for general and flag officer requirements/positions?
- How is information gathered and validated concerning the competencies needed (the "job description") for the position?
- How is individual officer development considered when establishing reserve component positions?
- What characteristics are considered in establishing the position as active or reserve component?
- Is there an understood boundary between characteristics for general and flag officers and SES positions?
- Is there a process for reviewing, updating, and capturing emerging requirements?

Managing Existing Positions, Authorizations, and Personnel

- When reserve component requirements are greater than reserve component authorizations ("headspace"), how do the services determine which positions to fill?
  - What are the alternate methods of filling positions and how are these chosen?
- How do the services determine which active component positions are to be filled by RC G/FOs?
- To what extent is there a reliance on exemptions for filling RC G/FO requirements? Should there be additional exemptions?
- How is individual officer development considered when assigning RC G/FOs to fill positions?

Specific Data Elements for Current General and Flag Officer Positions

- Grade (including previous increases or decreases in grade)
- Service and reserve component
- Joint service positions
- Full-time/part-time
- Competencies, experiences, or certifications required for position
- Overall functional area/line of operation (operations, headquarters, service support)
- Section of law from which authorized strength is earned
- Contingency or permanent position
- Command or noncommand position
- Other data elements as needed
- Any additional information that can be provided to help understand general and flag officer requirements processes and the management of RC G/FO positions and individuals.

APPENDIX C

Kapos Attributes

This appendix lists the 21 attributes used by Kapos in its 1996 review of Marine Corps general officer positions, as discussed in Chapter Four. Each position was evaluated to determine the presence or absence of those attributes, which were divided into four broad groups: nature of position, magnitude of responsibilities, significance of duties, and special qualifications.

Nature of Position

1. Level or echelon of the military establishment at which duty is performed
2. Rank of the official to whom the position reports
3. Rank of the majority of lateral counterparts
4. Span of control (rank, number, and diversity of direct subordinates)
5. Special authority that goes with the position by legislation or regulation
6. Independent decisionmaking authority.

Magnitude of Responsibilities

7. Number of personnel and commands under the position
8. Operating budget of command, including subordinate commands
9. Other money managed, such as military construction funds
10. Other product throughput
11. Value of equipment controlled
12. Value of real estate in the form of land
13. Value of facilities such as buildings and runways
14. Value of inventories controlled.

Significance of Duties

15. Duties in the international arena entailing independent dealings with foreign audiences
16. Duties at the seat of government having an impact on national defense
17. Duties involving significant exposure to the public and media
18. Duties entailing representation before Congress
19. Duties entailing direct support to the operating forces.

Special Qualifications

20. Unusual breadth of experience required
21. Special depth of skill or training other than the primary combat specialty.

APPENDIX D

LMI Validation Factors

The 16 factors described in this appendix were used by LMI to validate general and flag officer requirements (GAO, 2004, pp. 29–31). They are divided into three groups: nature of the position, magnitude of responsibilities, and significance of actions and decisions.

Nature of the Position

1. Characteristics of function
   a. type (e.g., command, general or coordinating staff, special staff, manager, deputy, specialist)
   b. scope (e.g., operational command, training command, installation command, personnel management, officer personnel management, legal affairs, information)
   c. level of function (e.g., national, secretarial, service, theater, field command).
2. Grade and position of
   a. superior
   b. principal subordinates
   c. lateral points of coordination (relative position within the military or governmental structure within which the position's function is performed).
3. Supervision over position
   a. proximity (remoteness or closeness of supervision)
   b. degree (independence of operation).
4. Official relations with U.S. and foreign governmental officials and with the public
   a. nature (e.g., reports to, works for, keeps informed, provides liaison)
   b. extent (e.g., primary function, frequent requirement, continuous additional duty, occasional requirement)
   c. level of official relations with U.S. and foreign governmental officials and with the public (e.g., governmental department or agency, national or local government, civil organizations, industry, press, nongovernmental organizations [NGOs], private volunteer organizations [PVOs]).

5. Reflection of national emphasis and determination (relation of position to national objectives and programs, special conditions under which the position was first established, or other reasons why the position reflects national will).
6. Special qualifications required by the position (any special qualifications—such as advanced education, particular training, or experience—that are essential to the proper execution of positional responsibilities).

Magnitude of Responsibilities

7. Mission(s) of organization and the special requirements of the position as it relates to the mission(s). (The nature of the responsibilities that are associated with the position and the need for multidimensional "executive skills." The mission of the organization is the key, day-to-day activities that are accomplished.)
8. Number, type, and value of resources managed and employed. Data should be displayed within three categories: operational control, administrative control, and immediate staff within each subsection
   a. military forces (number and type of forces normally assigned or programmed for planned or special operations)
   b. personnel (number of personnel by officer and warrant officer, enlisted, and civilian)
   c. value of equipment and properties (total value of equipment, supplies, and real property displayed in millions)
   d. total obligation authority
   e. foreign resources (scope and type of foreign resources involved, if any)
   f. other important resources.
9. Geographical area of responsibilities (the size, location, and, if appropriate, the criticality of the land, sea, or air spaces involved).
10. Authority to make decisions and commit resources (the scope of the position with respect to specific authority delegated to or withheld from the position in either routine or emergency situations).
11. Development of policy (involvement in the development of policy within the specific functional areas associated with the position, e.g., budget, program, communications, or manpower).
12. National commitment to international agreements (authority to make commitments to foreign nations or involvement in negotiating such commitments for the United States).
13. Auxiliary (supporting) authorities and responsibilities inherent in the position (inherent requirements charged to the position by virtue of situation, location, proximity, tradition, etc.).

Significance of Actions and Decisions

14. Impact on national security or other national interests (effect of mission accomplishment or position performance on the protection of national interests or the advancement of national programs).
15. Importance to current and future effectiveness and efficiency of the national defense establishment (effect on the force structure, operational capabilities, status of combat readiness, quality of personnel and equipment, cost effectiveness, command and control means, management procedures and techniques, responsiveness to national needs, or other factors).
16. Effect on the prestige of the nation or the armed forces (how effectiveness or accomplishment reflects on the stature of the nation and its armed forces, and influences the credibility of national aims and capabilities).

APPENDIX E

RAND Data Collection Protocol: Instrument

General Information

1. Duty title *(example: commander, director, special assistant, G3, etc.).*
2. Authorized grade *(O-7 to O-10).*
3. If an O-7 or above is not currently assigned to (filling) this billet,
   a. Approximate date on which the O-7 or above requirement was established (YYYYMMDD), exact date not required, year and month is sufficient.
   b. b.1. Since September 11, 2001, or going back as far as available records permit, has an O-7 or above at any time been assigned to (filled) this billet?
      ☐ Data not available
      ☐ Never
      ☐ Yes; provide dates____________________ *(YYYYMMDD to YYYYMMDD)*

      b.2. If currently filled by a G/FO, please provide the approximate dates (YYYYMMDD to YYYYMMDD)
4. Grade of incumbent
5. Organization name (please identify at the lowest organizational level appropriate, not the parent unit) *(e.g., for the commander of a unit, it is his unit, not the parent unit)*
6. Unit Identification Code (UIC)
7. Location *(of the immediate/lowest organization, later question will address area of responsibility/area supported).*
8. Dual-hatted or multi-hatted position (*this includes holding two different positions, as well as a second "title" with corresponding functions, but not a separate billet. Explain both functions (hats) in the same line because they are filled by one person).*
   a. Yes/No
   b. Identify grade and title of other position(s)

Background

9. What is the mission of the organization?
   Please provide the explicit stated mission of the organization in which the billet exists (not the parent organization, e.g., "7th Civil Support Command provides trained and ready, forward-stationed Consequence Management Command and Control, Civil Support Team, Civil Affairs, Enabler capabilities, as directed by USAREUR. Rapidly deploys immediate response capability and provides Title 10 responsibilities for European-based units as directed by USAREUR.")
10. What is the job description of the billet? (Official Description) *(Please describe the primary functions of the job as listed in an official billet description or similar source.)*
11. Was this billet created before or after September 11, 2001?
    - ☐ Before 11 Sept 2001
    - ☐ After 11 Sept 2001
    - ☐ Data not available/unknown
12. Since September 11, 2001, or going back as far as available records permit, has this billet been generally filled by reservist in which of the below statuses:
    - ☐ Individual Mobilization Augmentee (IMA)
    - ☐ Active Guard Reserve (AGR)/Full Time Support (FTS) and other categories of Full-Time Positions
    - ☐ Troop Program Units (TPU)/Selected Reserve (SELRES)/Traditional Reservist (TR)/Selected Marine Corps Reserve (SMCR)
13. The billet is currently filled under which statutory authority?
14. Since September 11, 2001, or going back as far as available records permit, approximately what percentage of the time has the position been filled by an officer of a higher or lesser grade, or an SES since September 11, 2001?
15. Since September 11, 2001, or going back as far as available records permit, has the position's higher headquarters realigned in a manner that realigned the position or shifted the position from AC to RC?
    a. Yes/No. If yes, to the extent that data is available:
    b. When? *(YYYYMMDD)*
    c. What was the nature of the realignment?
    d. Why was the realignment done?
16. To the extent that data is available, since September 11, 2001, was this position split out from another position?
    a. Yes/No. If yes:
    b. What position did this one originate from?
    c. What other positions were split out of it? *(Please focus on what functions/positions may need to be codified in authorizations.)*
17. If not captured above, please describe in broad terms the changes (if any) in the duties and responsibilities of the position that have occurred as a result of the demands on the force resulting from OIF, OEF, and other post–September 11, 2001 operations.

Nature of the Responsibilities

18. Characteristics of the organization
For example, 7th Civil Support Command would be an operational unit. Please choose from the dropdown choices. If "other" is chosen, please identify the organization in the designated box.
This position is in which type of organization? (Choose one)
☐ National *(e.g., NSC, National-level intelligence agencies)*
☐ OSD
☐ Service Secretariat *(e.g., SECARMY, SECNAV, SECAF)*
☐ Joint Staff
☐ Combatant Command *(e.g., CENTCOM, PACOM, CYBERCOM)*
☐ Service Chief Staff Headquarters *(e.g., OPNAV, HQDA, Air Staff)*
☐ State-level Command/Headquarters
☐ Major Service Headquarters *(e.g., TRADOC, ACC, COMNAVAIRFOR)*
☐ Theater *(e.g., ISAF, USFOR-A, NATO, CJTF-HOA)*
☐ Service Component Command *(e.g., ARCENT, NAVCENT, AFCENT)*
☐ Operational Unit *(e.g., Division Commander, Wing Commander)*
☐ Defense, Joint, or Service Agency *(e.g., DISA, JPRA)*
☐ Other

The following choices were added after the data call was collected
☐ Support Unit
☐ Training Unit

19. Characteristics of billet function
The following two questions establish the type and role of the position. For example, 7th Civil Support Commander would choose commander for part (a) and "Other: Civil Affairs" for part (b). If other is chosen, please identify the nature and/or role of the position in the designated box.
What is the nature of the billet position?
☐ Commander
☐ Deputy or Vice Commander
☐ Director
☐ Deputy Director
☐ Chief of Staff
☐ General/Flag Staff (e.g., J-1, J-2, J-3, or service equivalents)
☐ Program Manager
☐ Special Assistant/Mobilization Assistant
☐ Other

b. What is the role of the position? *(For Commanders/Directors this should be the same as the role of the organization. For other positions identify the specific function or role of the position within the organization.)*

- ☐ Administration
- ☐ Intelligence
- ☐ Operations
- ☐ Materiel and Logistics
- ☐ C4I
- ☐ Force Management, Development, Education and Training
- ☐ Strategic Plans and Policy
- ☐ State-specific functions
- ☐ Acquisition
- ☐ Research and Development
- ☐ Legal Affairs
- ☐ Congressional Affairs
- ☐ Medical
- ☐ Public Affairs
- ☐ Financial Management
- ☐ Other

The following choices were added after the data call was collected

- ☐ Civil Affairs
- ☐ Military Police
- ☐ Engineer
- ☐ Chaplain
- ☐ Support

20. Position of the billet within the organization.
    a. What is the grade and position of the position's superior(s)?
    b. What are the grades and positions of the position's principal subordinates? *For example, a Division/Wing commander would list subordinate Brigades/Squadrons; a Chief of Staff would list the 1-digit codes he manages. (Please keep answers to immediate superiors and principal subordinates. There may be instances where "none" is an appropriate answer.)*

21. Billet's role at the national and international level *(Please complete all boxes, even if they are all "N/A.")*

	Nature of Engagement with Each Type of Audience				
Audience	Function/Duty (i.e., 50% or more of work)	Frequent Function/Duty (i.e., 25–50% of work)	Occasional Additional Duty (i.e., 5–25% of work)	Rarely/Never (i.e., less than 5% of work)	N/A
a. Senior DoD (SD/DSD/USD, Service Secretary & Under Secretary level)					
b. Senior IC (DNI, Agency Director level)					
c. White House, NSC					
d. Congress, congressional staff					
e. Non-DoD, non-IC Cabinet-level					
f. State governors					
g. State legislatures					
h. Local authorities					
i. Foreign government (ministry level or higher)					
j. Foreign militaries (senior leader level)					
k. Industry senior leadership regarding ACAT I or II programs					
l. Nongovernmental organization senior leadership					
m. Press, national level					

Magnitude of Responsibilities

22. Resources managed and employed
    a. Military units. *(Number and type of units normally assigned or programmed)*
    b. Personnel. *(Number of authorized personnel by commissioned officer, warrant officer, enlisted, and civilian)*
    c. Value of equipment and properties. *(Total value of equipment, supplies, and real property displayed in millions)*
    d. Total obligation authority.
    e. Other important resources.
    Please round to whole numbers. For personnel break out commissioned officers/warrant officers/enlisted/civilian. For value of equipment and total obligation authority please provide response in millions. (e.g. 7th Civil Support Command):
    - ☐ Military Units: 7th CSC: 15 units (including 5x O-6 commands); 21st TSC: 40 units
    - ☐ Personnel: 12,000; 7th CSC: 318/38/658/12; 21st TSC: 691/0/125/6807
    - ☐ Value of equipment: $21 million
    - ☐ Total Obligation Authority: $23 million
    - ☐ None

23. What is the geographic area of responsibility? *(Identify the land, air, or sea area of operations for the unit/organization. If, however, it is a support unit that deploys or supports units or activities in a particular theater, region, or set of countries, mention the relationship and the supported area or number of countries supported.)*
24. Identify the role the position has in specific areas of policy and doctrine development, be specific about the type of policy, doctrine, etc. *(e.g., budget, program, communications, manpower, supporting the "x" service, or "y" COCOM).*
25. Does the position have authority to negotiate commitments or international agreements with foreign nations on behalf of the United States? If *yes*, please articulate under what circumstances, if able to do so in an UNCLASSIFIED manner *(e.g., yes; supports CONPLAN 4299/4269).*
26. Are there any other unique attributes or authorities associated with this position that we should know about? *(Specifically identify why there is a need for a General or Flag Officer in this position, e.g., the position requires the coordination of 6 O-6 level commands, or 90% of meetings that the individual attends are GO/FO-level meetings requiring commensurate level participation).*

Role in General/Flag Officer Development

27. What special qualifications, if any, required by the position are essential to the proper execution of positional responsibilities? *Please specify if each qualification is required by statue, policy, precedence, or some other factor (fill all that apply and identify the particular education, training, experience, etc. that is required; if they do not apply enter "N/A").*
    a. Advanced Education
    b. Specific Training
    c. Prior Experience
    d. Previous Billet
    e. Other
28. What aspects of this position are key in developing incumbents for advancement to more-senior general and flag officer positions?
29. Does the position afford an opportunity to acquire significant joint experience?

Impact of Not Having a General/Flag Officer

30. Is this position currently filled by a general or flag officer? If *no*, please complete questions 31–32.
31. Does the individual currently filling the position have legal authority to carry out all duties required of the position? If *no*, please explain why.
32. Does the practice of assigning lower grade officers to the position affect the ability of the organization to perform its mission? If *yes*, please explain why, and address whether the assignment of promotable/selected O-6s does or does not mitigate the challenges.

Conclusion

33. Is there anything else specific to this billet that you wish to add that was not covered?

APPENDIX F

RAND Data Collection Protocol: Development, Deployment, and Coding

Development of RAND Data Collection Protocol

The protocol for the data call (which is found in Appendix E) was primarily based on the LMI 2003 study that validated general and flag officer requirements. RAND researchers refined the LMI protocol to incorporate the lessons learned and address factors of particular relevance to the reserve components. RAND's data collection questionnaire contained four main sections: nature of the position, magnitude of responsibilities, role in general and flag officer development, and impact of not having a general or flag officer. The sections on nature of the position and magnitude of responsibilities contained similar wording to the LMI study, with some adaptation. The other two sections covered new material and were not contained in the LMI study. These were included to address deficiencies in the LMI study and specifically address the developmental value of general and flag officer positions—the latter having been identified early in the RAND analysis as a particularly important issue for the reserve general and flag officer community.

A general information section at the beginning of the data call obtained basic information about the position. This included title, grade, location, and other identifiers. Of note, RAND researchers included a question explicitly asking if the position is dual-hatted. Previous studies had been criticized for failing to collect these data. This basic information also helped RAND researchers eliminate double counting of positions.

The purpose of the background section was to determine the following position information:

- mission, history, and geographic context of the position
- precedent for alternate ranks/SES
- history of the evolution of the command and the position.

This background information provided important contextual information and organizational trends that helped interpret the remaining data correctly and apply criteria with an understanding of the overall context. We also determined whether the position had been created pre- or post-9/11, to help assess recent changes to fulfill emerging needs and capabilities as a result of the wars in Iraq and Afghanistan or the broader global counterterrorism campaign.

The inclusion of the section on nature of the responsibilities served to characterize the role and function of the position. Questions 18 and 19 were similar to questions that had been asked during the LMI study. However, where the LMI asked for "type," "scope," and "level of function," the RAND data set asked for "type of organization," "nature of the billet position," and "role of the position," which were included in an attempt to clarify what was being asked; a set of options were provided to further clarify the information desired. The categories in questions 18 and 19 were derived from the LMI data set with modifications.[1]

Question 20 was taken from the LMI data call as is, but only asked about the position's superiors and subordinates. Question 21 was based on LMI's question 4. RAND researchers believed it was important to include a section that addressed the external components of a job, as this was a limitation of the Hay methodology and had been highlighted as a factor that was particularly important in identifying positions requiring general or flag officers. Therefore, RAND researchers expanded the question to include all actors deemed of a high level that would necessitate a general or flag officer for interactions. To ensure consistent responses, RAND researchers provided a limited set of options based on how frequently the interactions occurred. Because reserve officers often act in a part-time capacity, RAND researchers used percentage of time as a measurement of frequency, as follows:

- primary function/duty (i.e., 50 percent or more of work)
- frequent function/duty (i.e., 25–50 percent of work)
- occasional additional duty (i.e., 5–25 percent of work)
- rarely/never (i.e., less than 5 percent of work)
- N/A.

The categories included for interaction included:

- senior DoD (at the secretary, deputy secretary, under secretary, service secretary, and service under secretary levels)

[1] For instance, a high number of inputs in the LMI data set were classified as "other." A thorough review of the "other" inputs led to RAND researchers creating a more comprehensive selection of answers. They were intended to be as inclusive as possible, but did contain an "other" category to catch any anomalies. However, it became clear after submission by the services that the list of choices was still not sufficient. As part of the data review process, RAND researchers added a few more categories based on what was being written in for "other" and recoded the data. RAND shared the recoded data with action officers during working meetings. The information in questions 18 and19 was collected to determine the range of characteristics associated with positions of equivalent rank. These data were useful in helping to ensure consistency across the services for similar positions.

- senior Intelligence Community (Director of National Intelligence, and agency director level)
- White House and National Security Council
- Congress and congressional staff
- other U.S. government departments (at the secretary and under secretary levels)
- state governors and legislatures
- foreign governments and militaries (at the ministry level, senior military leader level, or higher)
- industry (senior industry leaders on issues regarding ACAT I or II programs)
- nongovernmental organizations (at the organization leader level)
- national-level press.

Aspects considered under the section on magnitude of responsibilities included both command responsibilities (authority over forces and personnel) and fiscal/financial responsibilities (assigned resources and ability to obligate the U.S. government). The collection of this information helped to determine the scope and magnitude of the responsibilities associated with each position. The LMI study contained a similar section and questions 22–25 were based on those questions included in their protocol. Of note, in question 22, RAND researchers eliminated the breakdown of resources managed by type of control (administrative, operational, immediate staff). This was intended to simplify the responses; however, it may have resulted in some inconsistency in how services interpreted the question. Some services interpreted the question rather narrowly in their original submission, but after working group meetings, resubmitted the data in better alignment with what other services had submitted. Question 26 was a new question added by RAND researchers to catch any authorities not covered by the previous questions.

The section on roles in general and flag officer development was new, and was included to understand the role each position plays in preparing general and flag officers for the highest levels of command and promotion to more-senior rank. The inclusion of this section arose out of early discussions with reserve component leaders who made it clear that positions were needed for sustaining professional development of general and flag officers, leading ultimately to the four-star Chief of the National Guard. The reserve component has fewer opportunities to develop the skills needed to promote up to 3- and 4-star levels, and identifying positions that would be key for general or flag officer development is therefore an important factor in deciding whether general or flag officer positions should or should not be eliminated.

The final section was included to determine the impact of workarounds. This topic was not included in the 2003 LMI study and was noted as a gap in the 2004 GAO report. Workarounds are the practice of assigning O-6s to general or flag officer positions to fill the gap between requirements and authorizations. However, it is not always clear whether an O-6 filling a general or flag officer position has all appropriate authorities. Question 31, therefore, was included in the protocol to help determine whether the position has all the required authorities if it is not filled with a general or flag officer. Question 32 assesses the impact that assigning officers of less than general and flag officer grade has on the unit's mission.

Data Collection Process

After completion of writing the data call questions, RAND researchers conducted an internal trial of the protocol using position descriptions provided by the services. This allowed RAND researchers to validate (and as necessary refine) the questions. RAND researchers shared the data call questions with OASD/RA, military department, and Joint Staff representatives, and met with them to review the data call questions. The meeting resulted in some modifications and the revised data call was approved.

RAND researchers created a spreadsheet for data entry based on the data call protocol. An instruction sheet, a complete data call protocol in Microsoft Word document form, and an Excel spreadsheet were sent to OASD/RA for further dissemination to the reserve components and joint organizations. Each reserve component and joint organization was responsible for completing an input for its RC G/FO requirements. The Joint Staff was responsible for completing the spreadsheet for joint positions regardless of which service filled the position. This was done to avoid double-counting positions. All components were requested to provide data on all the existing requirements (as of February 2015) for RC G/FOs. This included any position that (a) requires a general or flag officer and (b) must be filled by a person coming from one of the reserve components. This last condition applies *regardless of the individual's status when incumbent in the position*. For example, the Chiefs of the Army, Navy, Marine Corps, and Air Force Reserves are all on active duty status and count against active duty headspace; however, by law they must be selected from the corresponding reserve components, so they are included in the data set.

Upon receipt of the spreadsheet, the parties had 30 days to complete it and return to OASD/RA. OASD/RA then forwarded all submissions to RAND. During the 30-day period, RAND researchers met individually with each service to assist with any questions or issues. During this working period, a few minor issues with the usability of the spreadsheet were identified and fixed.

Data Clean-Up, Quality Assurance, and Coding

After receiving all of the submissions, RAND researchers first conducted a comprehensive review of the position descriptions in an attempt to eliminate any duplicates. Submitted data were also compared with the Blue Book to ensure that joint positions were captured. Finally, a review was conducted of all submissions to ensure that all submissions had essential entries complete and answered in the provided format. During this review, a number of duplicates, missing positions, and incomplete entries were discovered. RAND researchers followed up with the respective reserve components to rectify these issues. RAND researchers also tried to ensure consistent reporting across services. For example, some reserve components included their O-9 positions while others did not; RAND researchers followed up to ensure that all reserve components and joint organizations reported these positions. Similarly, two of the service components did not consistently report positions that were not currently filled; RAND researchers worked with them to obtain data for the missing positions.

To ensure consistency across services, RAND researchers recoded questions 18 and 19. Additional choices were also added in order to better bin many of the positions that had been listed as "other." There was also a discrepancy as to how different services reported the nature of the position question when it came to the single digit J-, G-, N-, S-, and A- code positions on headquarters staffs (e.g., the J-1, J-2, J-3, . . . codes on the joint or combatant command staffs; the G-1, G-2, . . . codes on an Army staff). These positions, collectively referred to as "J-codes," are senior positions on the general staff of a major headquarters, but are often titled "director." For example the J-3 is also known as the director of operations. To ensure continuity across all the services and joint staff, RAND researchers coded these positions as general/flag staff and the director category was reserved for directors of independent organizations.

RAND researchers then developed an initial framework of factors required to necessitate a general or flag officer in a position. This initial criterion was based on factors that RAND researchers deemed necessary to validate a general or flag officer requirement. Reasons for the inclusion of each of these factors were discussed in Chapter Four. The factors and the corresponding data call question used to assess the filter factors employed in Chapter Four are as follows:

- TAG or AAG
  - Question 1: Duty Title
- Commander or deputy/vice commander
  - Question 19a: What is the nature of the position?
- Senior subordinates (general or flag officer, SES, O-6)
  - Question 20b: What are the grades and positions of the position's principal subordinates?
- Senior interactions (primary or frequent function)
  - Question 21a–m: Billet's role at the national and international level
- Role in setting policy
  - Question 24: Identify the role the position has in specific areas of policy development and for what organization
- Can obligate the United States

  - Question 25: Does the position have authority to negotiate commitments or international agreements with foreign nations on behalf of the United States?
- Director or deputy/vice director
  - Question 19a: What is the nature of the position?
- New requirement (since September 11, 2001)
  - Question 11a: Was this position created before or after September 11, 2001?
  - Question 15a: To the extent that data are available, since September 11, 2001, was this position split out from another position?

Each submission for each of these questions was then coded. Once RAND researchers completed the initial data analysis, they met both individually and collectively with the reserve components and the Joint Staff to discuss the initial findings and raise further questions about the data. This process allowed all services and the Joint Staff an opportunity to review their submissions and for RAND researchers to clarify some emerging questions. It also ensured that the services were interpreting and answering the questions in a consistent manner. As a result of these meetings, reserve components and the Joint Staff provided corrections to their submissions. These corrections also included several additional positions requiring RC G/FOs that we had identified as missing from the original submissions. These new data were incorporated into the data set and recoded. This provided the baseline of data used for all of the analysis in this study.

APPENDIX G

Results of a More Stringent Analysis of RC G/FO Requirements

Chapter Four describes the process RAND researchers used to review current RC G/FO requirements to determine whether some positions might be candidates for elimination, downgrading to a lower rank, or conversion to senior civilian positions. RAND researchers identified seven factors associated with a general or flag officer and filtered current requirements through those factors. Any position that did not have any of the key factors associated with a general or flag officer becomes a candidate for detailed review by the services to determine if in fact the position should be eliminated, downgraded, or converted. RAND researchers first applied the factors less stringently, so as to give the benefit of the doubt to accepting general and flag officer requirements. Then we applied the factors more stringently, which produced the 95 positions that are listed in Table G.1.

Table G.1
Ninety-Five Positions Remaining After Stringent Application of Factors Relating to RC G/FO Positions

Duty Title	Organization	Location	Positions Identified with Less Stringent Factors
Air National Guard			
Special assistant to the director, Air National Guard for International Affairs	Director, Air National Guard	Arlington, Virginia	X
Assistant to the commander	U.S. Air Forces Central Command	Shaw Air Force Base, South Carolina	X
Assistant to the Surgeon General of the Air Force	Air Force Surgeon General	Pentagon, Arlington, Virginia	X
Assistant to the commander	Air Mobility Command	Scott Air Force Base, Illinois	X
Assistant to the commander	Air Education and Training Command	Joint Base San Antonio, Texas	X

Table G.1—Continued

Duty Title	Organization	Location	Positions Identified with Less Stringent Factors
Assistant to the commander	U.S. Air Forces in Europe	Ramstein Air Base, Germany	X
Assistant to the commander	Pacific Air Forces	Hickam Air Force Base, Hawaii	X
Assistant to the commander	Air Force Special Operations Command	Hurlburt Field, Florida	X
Assistant to the Command Surgeon	Air Combat Command, Surgeon General	Langley Air Force Base, Virginia	X
Assistant to the Command Surgeon	Air Mobility Command, Surgeon General	Scott Air Force Base, Illinois	X
Assistant to the commander	14th Air Force	Vandenberg Air Force Base, California	X
Assistant to the commander	18th Air Force	Scott Air Force Base, Illinois	X
Assistant to the director, Logistics, Installations and Mission Support, A4	Director, Air National Guard	Joint Base Andrews, Maryland	X
Assistant to the commander	U.S. Transportation Command	Scott Air Force Base, Illinois	X
Assistant to the Assistant Secretary of the Air Force, Manpower and Reserve Affairs	Office of the Assistant Secretary of the Air Force for Manpower and Reserve Affairs	Pentagon, Arlington, Virginia	X
Assistant to the Chief of Chaplains	Air Force Chief of Chaplains	Bolling Air Force Base, Washington, D.C.	
Director, Operations, Plans and Requirements, A3	Director, Air National Guard	Arlington, Virginia	
Special assistant to the director	Director, Air National Guard	Arlington, Virginia	
Director, Strategic Plans and Programs, A8	Director, Air National Guard	Arlington, Virginia	
Assistant to the commander	Air Force Space Command	Peterson Air Force Base, Colorado	
Assistant to the commander	Air Combat Command	Langley Air Force Base, Virginia	
Assistant to the Judge Advocate General of the Air Force	Office of the Air Force Judge Advocate General	Washington, D.C.	
Assistant to the Deputy Judge Advocate General of the Air Force	Office of the Air Force Judge Advocate General	Washington, D.C.	
Assistant to the Chief Nurse of the Air Force	Office of the Air Force Chief Nurse	Washington, D.C.	

Table G.1—Continued

Duty Title	Organization	Location	Positions Identified with Less Stringent Factors
Assistant to the commander	1st Air Force	Tyndall Air Force Base, Florida	
Assistant to the commander	24th Air Force	Lackland Air Force Base, Texas	
Assistant to the director, Manpower, Personnel and Services, A1	Director, Air National Guard	Arlington, Virginia	
Assistant to the director, Intelligence, Surveillance and Reconnaissance, A2	Director, Air National Guard	Joint Base Andrews, Maryland	
Assistant to the director, Strategic Plans and Programs, A5	Director, Air National Guard	Arlington, Virginia	
Assistant to the director, Information Dominance, A6	Director, Air National Guard	Joint Base Andrews, Maryland	
Assistant to the director, Installations and Mission Support, A7	Director, Air National Guard	Joint Base Andrews, Maryland	
Assistant to the commander	Air Force Material Command	Wright-Patterson Air Force Base, Ohio	
Assistant to the director of operations	Air Mobility Command	Scott Air Force Base, Illinois	
Army National Guard			
Army National Guard assistant to the Judge Advocate General	Army Judge Advocate General	Washington, D.C.	X
Deputy Chief of Staff	Army Training and Doctrine Command	Fort Monroe, Virginia	X
Army National Guard Deputy Chief of Staff and Assistant Surgeon General for Mobilization, Readiness and National Guard Affairs	Army Surgeon General	San Antonio, Texas	
Army National Guard assistant to the Chief of Chaplains	Headquarters, Department of the Army	Pentagon, Arlington, Virginia	
Army National Guard assistant for Reserve Affairs, Chief of Engineers	Corps of Engineers	Washington, D.C.	
Special assistant to the director	Director, Army National Guard	Arlington, Virginia	

Table G.1—Continued

Duty Title	Organization	Location	Positions Identified with Less Stringent Factors
Special assistant to the director	Director, Army National Guard	Arlington, Virginia	
Deputy Surgeon General for Army National Guard	Army Surgeon General	Washington, D.C.	
NGB Joint Staff			
National Guard assistant to the commander and liaison to the chief, NGB	U.S. Northern Command	Peterson Air Force Base, Colorado	
National Guard assistant to the commander	U.S. Cyber Command	Fort Meade, Maryland	
Director, Manpower and Personnel, J-1	NGB	Arlington, Virginia	
Director, Intelligence, J-2/6	NGB	Arlington, Virginia	
Director, Domestic Operations, J-3/7	NGB	Arlington, Virginia	
Deputy director, Domestic Operations, J-3/7	NGB	Arlington, Virginia	
Director, Logistics, J-4	NGB	Arlington, Virginia	
Deputy director, Strategic Plans and Policy, J-5	NGB	Arlington, Virginia	
Director, Force Structure, Resources and Assessment, J-8	NGB	Arlington, Virginia	
Special assistant to chief of the NGB for diversity	NGB	Pentagon, Arlington, Virginia	
Air Force Reserve			
Mobilization assistant to the commander	618th Air Operations Center (Tanker Airlift Control Center)	Scott Air Force Base, Illinois	X
Mobilization assistant to the commander	18th Air Force	Scott Air Force Base, Illinois	X
Mobilization assistant to the commander	U.S. Air Force Expeditionary Center	Joint Base McGuire-Dix-Lakehurst, New Jersey	X
Mobilization assistant to the commander	Air University	Maxwell Air Force Base, Alabama	X
Mobilization assistant to the commander	Curtis E. Lemay Center for Doctrine Development and Education	Maxwell Air Force Base, Alabama	X
Mobilization assistant to the commander	2nd Air Force	Keesler Air Force Base, Mississippi	X

Table G.1—Continued

Duty Title	Organization	Location	Positions Identified with Less Stringent Factors
Mobilization assistant to the commander	19th Air Force	Joint Base San Antonio, Texas	X
Mobilization assistant to the commander	8th Air Force	Barksdale Air Force Base, Louisiana	X
Mobilization assistant to the commander	20th Air Force	F. E. Warren Air Force Base, Wyoming	X
Army Reserve			
Deputy chief for Mobilization and Reserve Affairs	U.S. Army Europe and 7th Army	Campbell Barracks, Germany	X
General officer support	Office of the Chief, Army Reserve	Fort Belvoir, Virginia	
Director, Cybersecurity	Office of the Army Chief Information Officer, G-6	Washington, D.C.	
Assistant Surgeon General for Mobilization, Readiness and Army Reserve Affairs	Office of the Army Surgeon General	Falls Church, Virginia	
Assistant Chief of Chaplains for Mobilization and Readiness	Office of the Army Chief of Chaplains	Washington, D.C.	
Chief judge	U.S. Army Legal Services	Arlington, Virginia	
Chief of staff	U.S. Forces Command	Fort Bragg, North Carolina	
Assistant deputy chief of staff, G-4, Logistics for Mobilization and Training	Headquarters, Department of the Army	Washington, D.C.	
Assistant deputy chief of staff, G-3/5/7, Operations and Plans	Headquarters, Department of the Army	Pentagon, Arlington, Virginia	
Assistant deputy chief of staff, Personnel for Mobilization and Reserve Affairs	Headquarters, Department of the Army	Pentagon, Arlington, Virginia	
Chief of staff, Eighth Army	Headquarters, Eighth Army	U.S. Army Garrison, Yongsan	
Deputy chief of staff	Office of the Chief, Army Reserve	Fort Belvoir, Virginia	
Chief of staff	Army Training and Doctrine Command	Fort Eustis, Virginia	

Table G.1—Continued

Duty Title	Organization	Location	Positions Identified with Less Stringent Factors
Marine Corps Reserve			
Assistant deputy commandant for Aviation (Mobilization)	Headquarters, Marine Corps	Pentagon, Arlington, Virginia	
Assistant deputy commandant for Combat Development and Integration (Mobilization)	Marine Corps Combat Development Command	Marine Corps Base Quantico, Virginia	
Assistant deputy commandant for r Plans, Policy, and Operations (Mobilization)	Headquarters, Marine Corps	Pentagon, Arlington, Virginia	
Assistant deputy commandant for Programs and Resources (Mobilization)	Headquarters, Marine Corps	Pentagon, Arlington, Virginia	
Assistant deputy commandant for Installations and Logistics (Mobilization)	Headquarters, Marine Corps	Pentagon, Arlington, Virginia	
Navy Reserve			
Assistant Deputy Surgeon General For Reserve Affairs	Bureau of Medicine and Surgery	Falls Church, Virginia	
Deputy Chief of Chaplains for Reserve Matters	Office of the Chief of Chaplains	Washington, D.C.	
Reserve chief staff officer	Bureau of Medicine and Surgery	Falls Church, Virginia	
Reserve Chief Staff Officer for Surgeon General of the Navy	Bureau of Medicine and Surgery	Falls Church, Virginia	
Joint Staff and Joint Organizations			
Director, Defense Logistics Agency Joint Reserve Force, J-9	Defense Logistics Agency	Fort Belvoir, Virginia	
Reserve deputy director, Joint Doctrine, Training, and Force Development, J-7	Joint Staff	Suffolk, Hampton Roads Virginia	
Mobilization assistant	Defense Contract Management Agency	Fort Lee, Virginia	
Director of Logistics and Engineering, J-4	North American Aerospace Defense Command and U.S. Northern Command	Peterson Air Force Base, Colorado	
Mobilization assistant to the commander	North American Aerospace Defense Command	Peterson Air Force Base, Colorado	

Table G.1—Continued

Duty Title	Organization	Location	Positions Identified with Less Stringent Factors
Special assistant to the commander for National Guard Matters and Liaison for the Chief, NGB	North American Aerospace Defense Command and U.S. Northern Command	Peterson Air Force Base, Colorado	
Special assistant to the commander for Reserve Matters	North American Aerospace Defense Command and U.S. Northern Command	Peterson Air Force Base, Colorado	
Mobilization assistant to the director, Operations, J-3	North American Aerospace Defense Command	Peterson Air Force Base, Colorado	
Mobilization assistant to the director, Operations, J-3	Headquarters, U.S. Pacific Command	Camp Smith, Hawaii	
Mobilization assistant to the director, Logistics, Engineering, and Security Cooperation, J-4	Headquarters, U.S. Pacific Command	Camp Smith, Hawaii	
Mobilization assistant to the commander, Director, Joint Reserve Directorate, J-10, and Commander Joint Strategic Reserve Unit	Headquarters, U.S. Strategic Command	Offutt Air Force Base, Nebraska	
Mobilization assistant to the deputy commander	Headquarters, U.S. Strategic Command	Offutt Air Force Base, Nebraska	
Mobilization assistant to the director, Global Operations, J-3	U.S. Strategic Command	Offutt Air Force Base, Nebraska	

APPENDIX H

Empirical Analysis of Development Aspects of RC G/FO Positions

In Chapter Four, we presented a methodology to evaluate the validity of RC G/FO requirements. That analysis identified 95 positions that should be further reviewed to determine if they remain valid requirements. One justification for retaining these positions is their use for developmental purposes—that is, the positions are regularly used to prepare general and flag officers for positions of greater responsibility. RAND researchers evaluated these 95 positions to determine if they played a significant role in officer development. Thirty-four of the positions were ruled out because they are coded for O-8—the highest grade that most reserve component officers achieve and therefore a position unlikely to be used for developmental reasons. We then assessed the role of the remaining 61 positions in the career progression of general and flag officers, looking for evidence that these positions were being used by the service for developmental purposes. To do so, we examined the biographies of active and retired general and flag officers from each of the reserve components, who held these positions dating back to September 2001. Table H.1 contains the detailed results of this evaluation, which is summarized in Chapter Five.

Table H.1
Results of an Empirical Analysis of Development Aspects of RC G/FO Positions

Duty Title	Role in Career Progression for General and Flag Officers
Air National Guard	
Assistant to the Chief Nurse of the Air Force	Two officers have gone on to hold subsequent positions—one as an AAG—and both were later promoted to O-8.
Assistant to the Air Force Chief of Chaplains	One officer became an O-8 and the director of the National Guard Chaplain Service, and the other retired from the position.
Assistant to the director, Manpower, Personnel, and Services, A1	One officer was a dual-hatted AAG and is still active as an O-7, although not still in the position.
Assistant to the director, Intelligence, Surveillance, and Reconnaissance, A2	The incumbent is triple-hatted as an AAG and Air National Guard Assistant to the Commander of a combatant command. He was promoted to O-8 while in the A2 assignment.

Table H.1—Continued

Duty Title	Role in Career Progression for General and Flag Officers
Director, Operations, Plans, and Requirements, A3	The incumbent is a dual-hatted AAG. One other was also dual-hatted as Air National Guard Chief of Staff at the state level and retired as an O-7 after a position on the NGB Joint Staff.
Assistant to the director, Logistics, Installations, and Mission Support, A4	One officer was dual-hatted as Air National Guard Chief of Staff at the state level and is currently a state TAG.
Assistant to the director, Strategic Plans and Programs, A-5	The incumbent is a dual-hatted AAG. Two others were dual-hatted AAGs, both are now Air National Guard Assistants to the Commander of a major command, and both are O-8s.
Assistant to the director, Communications, A6	Three officers have been dual-hatted. One was also Director of the Joint Staff at the state level and another was also an AAG. Both retired as O-7s. One was also Chief of Staff at the state level and is now Assistant to the Commander, Air Force Space Command as an O-8.
Assistant to the director, Installations and Mission Support, A7	The incumbent is a dual-hatted AAG.
Director, Strategic Plans and Programs, A-8	The incumbent is a dual-hatted AAG. One other was also dual-hatted as AAG and is currently a state TAG.
Special assistant to the director	Two who held this position were promoted to O-8 while four others retired in the position. Neither of the two future O-8s served in the position for more than four months.
Assistant to the commander, U.S. Air Forces in Europe	Besides the incumbent, one officer was dual-hatted and eventually became a TAG, and one retired from the position.
Assistant to the commander, Pacific Air Forces	Besides the incumbent, one officer retired from the position.
Assistant to the commander, Air Force Special Operations Command	Besides the incumbent, one officer was dual-hatted as Air National Guard Chief of Staff at the state level, a position from which he retired.
Assistant to the Judge Advocate General of the Air Force	Besides the incumbent, two retired from the position.
Assistant to the Deputy Judge Advocate General of the Air Force	We found no match for this position in the Air National Guard biographies.
Assistant to the Command Surgeon, Air Combat Command	Besides the incumbent, one left the position to become Air National Guard Assistant to the Air Force Surgeon General for Operations and Policy, then returned to the position and finally served as Special assistant to the director, Air National Guard before retiring as an O-7.
Assistant to the Command Surgeon, Air Mobility Command	Besides the incumbent, one officer later became an AAG, and one retired from the position.
Assistant to the commander, 14th Air Force	Two officers have held the position while dual-hatted as AAGs. One retired from the position as an O-7 and the other is currently an O-8 and the Mobilization Assistant to the Commander, North American Aerospace Defense Command.
Assistant to the commander, 18th Air Force	We found no match for this position in the Air National Guard biographies, although one O-7 retired from a position listed as "Air National Guard Advisor to the Commander, 18th Air Force" in 2004.

Table H.1—Continued

Duty Title	Role in Career Progression for General and Flag Officers
Assistant to the commander, 24th Air Force	Two officers have held the position while dual-hatted. The incumbent is also the Director of Joint Staff at the state level; the other was dual-hatted as an AAG, has held two subsequent positions, and is now an O-8.
Assistant to the commander, Air Force Material Command	The incumbent is dual-hatted as Chief of Staff and Commander of a state National Guard, and another officer retired from the position.
Assistant to the director of operations, Air Mobility Command	The incumbent is dual-hatted as an AAG, and another officer was dual-hatted while Air National Guard Chief of Staff at the state level and is now an AAG.
Assistant to the commander, U.S. Transportation Command	The incumbent is dual-hatted as an AAG. Another O-7 is currently dual-hatted as an AAG and as "Air National Guard Advisor to the U.S. Transportation Command."
Assistant to the Assistant Secretary of the Air Force, Manpower and Reserve Affairs	Two officers have held the position while dual-hatted as AAGs. One retired from the position and the other became an O-8 and held one subsequent position.
Army National Guard	
Special assistant to the director (two individuals held this position at the time of our data collection)	Seven officers have held this position in recent years; at least four reached the grade of O-8, and the other three are all still active as O-7s. One became acting director of the Army National Guard, and another is the Deputy Commanding General of United States Army North/Fifth United States Army.
Assistant to the Army Chief of Chaplains	One officer retired from this position.
Assistant for Reserve Affairs, Corps of Engineers	The incumbent is dual-hatted as an AAG.
Deputy Surgeon General for Army National Guard	Two officers retired from the position, and the incumbent is dual-hatted as Director of the Office of the Joint Surgeon, NGB.
Assistant to the Judge Advocate General	One officer held the position for four years, including one year while dual-hatted, and is currently an AAG.
Deputy chief of staff and Assistant Surgeon General for Mobilization, Readiness and National Guard Affairs	One officer retired from the position, and the other is now Deputy Surgeon General for the Army.
NGB Joint Staff	
Director, Manpower and Personnel, J-1	Besides the incumbent, two officers retired from the position, and one was promoted to O-8, had several more assignments, then retired as a TAG.
Director, Intelligence, J-2/6	One officer retired from this position.
Deputy director, Domestic Operations, J-3/7	One officer retired from this position.
Director, Logistics, J-4	Besides the incumbent, one officer retired from the position, and another subsequently served as Special assistant to the director, Army National Guard. All three served in the position as O-8s, even though this is listed as an O-7 position.
Deputy director, Strategic Plans and Policy, J-5	We found no match for this position in the Army National Guard biographies.

Table H.1—Continued

Duty Title	Role in Career Progression for General and Flag Officers
Director, Force Structure, Resources and Assessment, J-8	Besides the incumbent, one officer retired from the position.
Special assistant to the chief of the NGB for Diversity	We found no match for this position in the Army National Guard biographies.
Air Force Reserve	
Mobilization assistant to the commander, 618th Air Operations Center (Tanker Airlift Control Center)	One officer is now an O-8 and has had three subsequent mobilization assistant positions (to two numbered Air Forces and the Air Force Reserve Command), one is Director for Air Force Plans, Programs, and Requirements, and one has had two subsequent mobilization assistant positions. All are still on active duty.
Mobilization assistant to the commander, Curtis E. Lemay Center for Doctrine Development and Education	Besides the incumbent, one officer is now an O-8 and has had several subsequent assignments, another is still active as an O-7 with several subsequent assignments, and one retired from the position.
Mobilization assistant to the commander, 2nd Air Force	Ten Air Force Reserve general officers have served in one of the numbered Air Force positions coded for an O-7, and at least five of the ten reached the grade of O-8 (eventually it may be more than five, since some O-7s are still on active duty).
Mobilization assistant to the commander, 19th Air Force	Ten Air Force Reserve general officers have served in one of the numbered Air Force positions coded for an O-7, and at least five of the ten reached the grade of O-8 (eventually it may be more than five, since some O-7s are still on active duty).
Mobilization assistant to the commander, 20th Air Force	Ten Air Force Reserve general officers have served in one of the numbered Air Force positions coded for an O-7, and at least five of the ten reached the grade of O-8 (eventually it may be more than five, since some O-7s are still on active duty).
Army Reserve	
Assistant Surgeon General for Mobilization, Readiness and Army Reserve Affairs	Besides the incumbent, two officers went on to command a medical brigade, and one retired from the position. None reached the grade of O-8.
Assistant Chief of Chaplains for Mobilization and Readiness	Besides the incumbent, two officers retired from this position.
Chief of staff, Eighth Army	Besides the incumbent, three officers retired from the position, and one commanded a regional support command afterward. Everyone prior to the incumbent served in the position as an O-8, so it appears to have been downgraded recently. Perhaps assignment patterns will change in the future.
Deputy chief of staff, Office of the Chief, Army Reserve	This appears to be new, with no record of officers having served in the position prior to the incumbent.

Table H.1—Continued

Duty Title	Role in Career Progression for General and Flag Officers
Marine Corps Reserve	
Assistant deputy commandant for Aviation (Mobilization)	Besides the incumbent, one officer retired from the position.
Assistant deputy commandant for Plans, Policy and Operations (Mobilization)	This position is vacant, and we found no record of any incumbent.
Assistant deputy commandant for Programs and Resources (Mobilization)	This position is vacant, and we found no record of any incumbent.
Assistant deputy commandant for Installations and Logistics (Mobilization)	This position is vacant, and we found no record of any incumbent.
Navy Reserve	
Deputy Chief of Chaplains for Reserve Matters	Four officers have retired from this position.
Reserve chief staff officer, Bureau of Medicine and Surgery	We found no match for this position in the Navy biographies.
Joint Staff	
Reserve deputy director, Joint Doctrine, Training, and Force Development, J-7 Joint Staff	The incumbent is an Air Force Reserve O-7.
Mobilization assistant, Defense Contract Management Agency	The incumbent is an Air Force Reserve O-7. A second Air Force Reserve O-7 is now a mobilization assistant at Air Force Space Command.
Mobilization assistant to the director for operations, J-3, U.S. Pacific Command	A Navy Reserve O-7 is the incumbent. An Air Force Reserve O-7 is now Mobilization Assistant to the Vice Commander, Pacific Air Forces.
Mobilization assistant to the Directorate of Logistics, J-4, U.S. Pacific Command	The incumbent is an Army National Guard O-7, and another Army National Guard officer retired from the position.
Mobilization assistant to the director, Global Operations J-3, U.S. Strategic Command	The incumbent is an Air Force Reserve O-7. One Air Force Reserve officer has held two subsequent positions at U.S. Strategic Command and is now the Mobilization Assistant to the Deputy Commander, U.S. Strategic Command as an O-8. Another Air Force Reserve O-7 is now the Mobilization Assistant to the Commander, 14th Air Force.
Director of Logistics and Engineering, J-4, North American Aerospace Defense Command and U.S. Northern Command	The incumbent is a Regular Air Force O-6. An Army National Guard O-7 is still active in another position. An Air National Guard O-7 is now an AAG. Both are still O-7s. A Regular Marine Corps O-7 also served in the position.
Mobility assistant to the director of operations, J-3, North American Aerospace Defense Command	The incumbent is an Air Force Reserve O-7. Another Air Force Reserve officer has held two subsequent positions and is now Commander, 10th Air Force as an O-8.

Abbreviations

9/11	September 11, 2001, terrorist attacks
AAG	assistant adjutant general
ACAT	acquisition category
CJCS	Chairman of the Joint Chiefs of Staff
CJTF-HOA	Combined Joint Task Force–Horn of Africa
DISA	Defense Information Systems Agency
DISES	Defense Intelligence Senior Executive Service
DISL	Defense Intelligence Senior Level
DoD	U.S. Department of Defense
FY	fiscal year
GAO	General Accounting Office; Government Accountability Office
GOPSG	General Officer Position Study Group
IMA	individual mobilization augmentee
ISAF	International Security Assistance Force
JPRA	Joint Personnel Recovery Agency
NGB	National Guard Bureau
OASD/RA	Office of the Assistant Secretary of Defense for Reserve Affairs
OEPM	Officer and Enlisted Personnel Management
OSD	Office of the Secretary of Defense
OUSD(P&R)	Office of the Under Secretary of Defense for Personnel and Readiness
RC G/FO	reserve component general and flag officer
SES	Senior Executive Service
TAG	the adjutant general
USAFOR-A	U.S. Forces Afghanistan
USAREUR	U.S. Army in Europe
U.S.C.	United States Code

Bibliography

Acting Assistant Secretary of Defense for Force Management Policy, "Reserve Component (RC) General Flag (GO/FO) Authorizations," memorandum to the secretary of defense and deputy secretary of defense, Washington, D.C., February 20, 2001, not available to the general public.

Air Force Reserve 2014 General Officer Position Study Group, "Air Force Reserve General Officer Management Office," memorandum, May 10, 2014.

Campion, Michael A., Lisa Cheraskin, and Michael J. Stevens, "Career-Related Antecedents and Outcomes of Job Rotation," *The Academy of Management Journal*, Vol. 37, No. 6, December 1994, pp. 1518–1542.

Chairman of the Joint Chiefs of Staff, *CJSC Vision for Joint Officer Development*, Washington, D.C., November 2005. As of June 15, 2015:
http://www.dtic.mil/doctrine/education/officer_JPME/cjcsvision_jod.pdf

———, *Joint Officer Management Program Procedures*, CJCSI 1330.05, Washington, D.C.: Joint Staff, May 1, 2008. As of April 22, 2015:
http://www.dtic.mil/cjcs_directives/cdata/unlimit/1330_05.pdf

———, *Manpower and Personnel Actions Involving General and Flag Officers*, CJCSI 1331.01D, Washington, D.C.: Joint Staff, August 1, 2010. As of April 22, 2015:
http://www.dtic.mil/cjcs_directives/cdata/unlimit/1331_01.pdf

CJCS—*See* Chairman of the Joint Chiefs of Staff.

Conley, Raymond E., Ralph Masi, Bernard D. Rostker, Herbert J. Shukiar, and Steve Drezner, *Enhancing the Performance of Senior Department of Defense Civilian Executives, Reserve Component General/Flag Officers, and Senior Noncommissioned Officers in Joint Matters*, Santa Monica, Calif.: RAND Corporation, MG-621-OSD, 2008. As of April 22, 2015:
http://www.rand.org/pubs/monographs/MG621.html

Dean, James W., and Jim Shanley, *Learning from Experience: The Missing Link in Executive Development*, Chapel Hill, N.C.: University of North Carolina, Kenan-Flagler Business School, August 2006. As of April 22, 2015:
https://www.kenan-flagler.unc.edu/~/media/Files/documents/executive-development/learning-from-experience.pdf

Defense Manpower Requirement Reports, Office of the Under Secretary for Personnel and Readiness website, undated. As of May 22, 2015:
http://prhome.defense.gov/RFM/TFPRQ/Reports.aspx

Department of the Air Force, "Individual Reservist (IR) Management," Air Force Instruction 36-2629, Washington, D.C., August 13, 2012. As of July 27, 2015:
http://static.e-publishing.af.mil/production/1/af_a1/publication/afi36-2629/afi36-2629.pdf

Department of the Army, Reserve Component General Officer Personnel Management, Army Regulation 135-156, Washington, D.C.: Headquarters, May 17, 2007.

———, *Individual Mobilization Augmentation (IMA) Program*, Army Regulation 140-145, Washington, D.C.: Headquarters, July 17, 2012. As of July 27, 2015:
http://armypubs.army.mil/epubs/pdf/R140_145.PDF

DoD—*See* U.S. Department of Defense.

GAO—*See* U.S. Government Accountability Office.

GOPSG—*See* Air Force Reserve 2014 General Officer Position Study Group.

Gortney, William E., *General/Flag Officer Zero Based Review*, memorandum, Washington, D.C.: Joint Staff, August 3, 2010.

Hay Group, *Study of General/Flag Officer Requirements and Distributions in the Department of Defense*, Washington, D.C., 1988.

Jacobs, Jeffrey, *The Future of the Citizen-Soldier Force: Issues and Answers*, Louisville, Ky.: University Press of Kentucky, 2014.

Jacobsen, Don, "Making the Most of Developmental Assignments: Q&A with Author Cynthia McCauley," GovLeaders.org, 2007. As of April 22, 2015:
http://govleaders.org/development.htm

Kapos Associates, *Analysis of U.S. Marine Corps Reserve General Officer Billet Requirements*, Final Report, KAI-153-96F, Arlington, Va., August 31, 1996.

Kester, John G., "State Governors and the Federal National Guard," *Harvard Journal of Law and Public Policy*, Vol. 11, No. 1, 1988, pp. 177–212.

McCall, Morgan W., "Leadership Development Through Experience," *The Academy of Management Executive (1993-2005)*, Vol. 18, No. 3, August 2004, p. 127–130.

McCall, Morgan W., Michael M. Lombardo, and Ann M. Morrison, *The Lessons of Experience: How Successful Executives Develop on the Job*, New York: Lexington Books, 1988.

McCauley, Cynthia D., and Stèfane Brutus, *Management Development Through Job Experiences: An Annotated Bibliography*, Greensboro, N.C.: Center for Creative Leadership, 1998.

McGuire, Mark A., "Senior Officers and Strategic Leader Development," *Joint Force Quarterly*, Autumn/Winter 2001–2002, pp. 91–96.

Offenhauer, Priscilla, *General and Flag Officer Authorizations for the Active and Reserve Components: A Comparative and Historical Analysis*, Washington, D.C.: Federal Research Division, Library of Congress, December 2007.

Office of the Under Secretary of Defense for Personnel and Readiness, "OEPM Mission," web page, undated. As of June 4, 2015:
http://prhome.defense.gov/RFM/MPP/OEPM/Mission.aspx

———, *Reserve Component General Officer Strength Ceiling Exemption Plan*, memorandum for Assistant Secretary of the Army (Manpower and Reserve Affairs), December 2, 2003.

OUSD(P&R)—*See* Office of the Under Secretary of Defense for Personnel and Readiness.

Public Law 381, Officer Personnel Act of 1947, August 7, 1947. As of April 22, 2015:
https://bulk.resource.org/gao.gov/80-381/00001E22.pdf

Public Law 85-861, Sec. 3218, Reserves: Strength in Grade; General Officers in Active Status, September 2, 1958. As of April 22, 2015:
https://www.govtrack.us/congress/bills/85/hr8943/text

Public Law 96-513, Defense Officer Personnel Management Act, December 12, 1980. As of April 22, 2015:
http://www.gpo.gov/fdsys/pkg/STATUTE-94/pdf/STATUTE-94-Pg2835.pdf

Public Law 103-337, National Defense Authorization Act for Fiscal Year 1995, Title XVI Reserve Officer Personnel Management Act (ROPMA), October 5, 1994. As of April 22, 2015:
https://www.govtrack.us/congress/bills/103/s2182/text

Public Law 113–66, National Defense Authorization Act for Fiscal Year 2014, Sec. 514 Review of Requirements and Authorizations for Reserve Component General and Flag Officers in an Active Status, December 2013. As of April 22, 2015:
http://www.gpo.gov/fdsys/pkg/CPRT-113HPRT86280/pdf/CPRT-113HPRT86280.pdf

Reserve Officer Personnel Management Act, H.R. 1040 EH, May 13, 1993. As of April 22, 2015:
http://www.gpo.gov/fdsys/pkg/BILLS-103hr1040eh/pdf/BILLS-103hr1040eh.pdf

Rostker, Bernard D., Harry J. Thie, James L. Lacy, Jennifer H. Kawata, and Susanna W. Purnell, *The Defense Officer Personnel Management Act of 1980: A Retrospective Assessment*, Santa Monica, Calif.: RAND Corporation, R-4246-FMP, 1993. As of April 22, 2015:
http://www.rand.org/pubs/reports/R4246.html

Rousseau, Major Thomas L., *National Guard Combat Divisions: State Militia or Federal Muscle Force Structure Mix*, Fort Leavenworth, Kan.: School of Advanced Military Studies, U.S. Army Command and General Staff College, 1997. As of May 27, 2015:
http://www.dtic.mil/dtic/tr/fulltext/u2/a331391.pdf

Schirmer, Peter, *Computer Simulation of General and Flag Officer Management: Model Description and Results*, Santa Monica, Calif.: RAND Corporation, TR-702-OSD, 2009. As of June 4, 2015:
http://www.rand.org/pubs/technical_reports/TR702.html

Thie, Harry J., Margaret C. Harrell, Clifford M. Graf II, and Jerry M. Sollinger, *General and Flag Officer Careers: Consequences of Increased Tenure*, Santa Monica, Calif.: RAND Corporation, MR-868-OSD, 2001. As of April 22, 2015:
http://www.rand.org/pubs/monograph_reports/MR868.html

United States Code, Title 10, Armed Forces, January 7, 2011. As of April 22, 2015:
https://www.law.cornell.edu/uscode/text/10

———, Section 101, Definitions, January 7, 2011.

———, Section 525, Distribution of Commissioned Officers on Active Duty in General Officer and Flag Officer Grades, January 7, 2011.

———, Section 526, Authorized Strength: General and Flag Officers on Active Duty, January 7, 2011.

———, Section 527, Authority to Suspend Sections 523, 525, and 526, January 7, 2011.

———, Section 528, Officers Serving in Certain Intelligence Positions: Military Status, Application of Distribution and Strength Limitations; Pay and Allowances, January 7, 2011.

———, Section 3038, Office of Army Reserve: Appointment of Chief, January 7, 2011.

———, Section 5143, Office of Navy Reserve: Appointment of Chief, January 7, 2011.

———, Section 5144, Office of Marine Forces Reserve: Appointment of Chief, January 7, 2011.

———, Section 8038, Office of Air Force Reserve: Appointment of Chief, January 7, 2011.

———, Section 10101, Reserve Components Named, January 7, 2011.

———, Section 12001, Authorized Strengths: Reserve Components, January 7, 2011.

———, Section 12002, Authorized Strengths: Army and Air Force Reserve Components, Exclusive of Members on Active Duty, January 7, 2011.

———, Section 12003, Authorized Strengths: Commissioned Officers in an Active Status, January 7, 2011.

———, Section 12004, Strength in Grade: Reserve General and Flag Officers in an Active Status, January 7, 2011.

———, Section 12006, Strength Limitations: Authority to Waive in Time of War or National Emergency, January 7, 2011.

———, Section 14002, Reserve Active-Status Lists: Requirement for each Armed Force, January 7, 2011.

United States Code, Title 32, Armed Forces, January 7, 2011. As of April 22, 2015:
https://www.law.cornell.edu/uscode/text/32

U.S.C.—*See* United States Code.

U.S. Department of Defense, *Continuation of Commissioned Officers on Active Duty and on the Reserve-Active Status List*, DoDI 1320.08, Washington, D.C., March 14, 2007. As of April 22, 2015:
http://www.dtic.mil/whs/directives/corres/pdf/132008p.pdf

———, *Frocking of Commissioned Officers*, DoDI 1334.02, Washington, D.C., December 7, 2012. As of April 22, 2015:
http://www.dtic.mil/whs/directives/corres/pdf/133402p.pdf

———, *Special Selection Boards*, DoDI 1300.12, Washington, D.C., February 12, 2013a. As of April 22, 2015:
http://www.dtic.mil/whs/directives/corres/pdf/132011p.pdf

———, *Rank and Seniority of Commissioned Officers*, DoDI 1310.01, Washington, D.C., August 23, 2013b. As of April 22, 2015:
http://www.dtic.mil/whs/directives/corres/pdf/131001p.pdf

———, *Service Credit for Commissioned Officers*, DoDI 1312.03, November 22, 2013c. As of April 22, 2015:
http://www.dtic.mil/whs/directives/corres/pdf/131203p.pdf

———, *Separation of Regular and Reserve Commissioned Officers*, DoDI 1332.30, Washington, D.C., November 25, 2013d. As of April 22, 2015:
http://www.dtic.mil/whs/directives/corres/pdf/133230p.pdf

———, *Commissioned Officer Promotion Program Procedures*, DoDI 1320.14, December 11, 2013e. As of April 22, 2015:
http://www.dtic.mil/whs/directives/corres/pdf/132014p.pdf

———, *Military Officer Actions Requiring Presidential, Secretary of Defense, or Under Secretary of Defense for Personnel and Readiness Approval or Senate Confirmation*, DoDI 1320.04, Washington, D.C., January 3, 2014a. As of April 22, 2015:
http://www.dtic.mil/whs/directives/corres/pdf/132004p.pdf

———, *DoD Joint Officer Management (JOM) Program*, DoDI 1300.19, Washington, D.C., March 4, 2014b. As of April 22, 2015:
http://www.dtic.mil/whs/directives/corres/pdf/130019p.pdf

———, *Commissioned Officer Promotion Reports (COPRs)*, DoDI 1320.13, Washington, D.C., October 30, 2014c. As of April 22, 2015:
http://www.dtic.mil/whs/directives/corres/pdf/132013p.pdf

———, *Utilization of Enlisted Aides (EAs) on Personal Staffs of General and Flag Officers*, DoDI 1315.09, Washington, D.C., March 6, 2015a. As of April 22, 2015:
http://www.dtic.mil/whs/directives/corres/pdf/131509p.pdf

———, *Original Appointment of Officers*, DoDI 1310.02, Washington, D.C., March 26, 2015b. As of April 22, 2015:
http://www.dtic.mil/whs/directives/corres/pdf/131002p.pdf

———, Management of Individual Mobilization Augmentees (IMAs), DoDI 1235.11, Washington, D.C., July 10, 2015c. As of July 27, 2015:
http://www.dtic.mil/whs/directives/corres/pdf/123511p.pdf

U.S. General Accounting Office, *Military Officers: Assessment of the 1988 Defense Officer Requirements Study*, GAO/NSIAD-88-146, April 1988.

———, *General and Flag Officers DOD's Draft Study Needs Adjustments*, statement of Mark E. Gebicke, Director, Military Operations and Capabilities Issues, National Security and International Affairs Division, testimony before the Subcommittee on Military Personnel, Committee on National Security, House of Representatives, GAO/T-NSIAD-97-122, April 8, 1997a.

———, *General and Flag Officers: Number Required Is Unclear Based on DOD's Draft Report*, GAO/NSIAD-97-160, June 1997b.

———, *Military Personnel: General and Flag Officer Requirements Are Unclear Based on DOD's 2003 Report to Congress*, GAO-04-488, April 2004.

U.S. Government Accountability Office, *Military Personnel: DOD Could Make Greater Use of Existing Legislative Authority to Manage General and Flag Officer Careers*, GAO-04-1003, September 2004.

U.S. Navy, *Flag Officer Study: A Comprehensive Review of Navy's Flag Officer Requirements*, January 29, 1997.

U.S. Senate, *General and Flag Officer Requirements: Hearing Before the Subcommittee on Personnel of the Committee on Armed Services United States Senate*, Senate Hearing 112-258, Washington, D.C., September 14, 2011.

White House, Executive Order No. 13223, September 14, 2001. *Federal Register*, Vol. 66, No. 181, p. 48201.